Molecular Methods in Developmental Biology

METHODS IN MOLECULAR BIOLOGY™

John M. Walker, SERIES EDITOR

145. **Bacterial Toxins:** *Methods and Protocols*: edited by *Otto Holst, 2000*
144. **Calpain Methods and Protocols**, edited by *John S. Elce, 2000*
143. **Protein Structure Prediction:** *Methods and Protocols*, edited by *David Webster, 2000*
142. **Transforming Growth Factor-Beta Protocols**, edited by *Philip H. Howe, 2000*
141. **Plant Hormone Protocols**, edited by *Jeremy A. Roberts and Gregory A. Tucker, 2000*
140. **Chaperonin Protocols**, edited by *Christine Schneider, 2000*
139. **Extracellular Matrix Protocols**, edited by *Charles Streuli and Michael Grant, 2000*
138. **Chemokine Protocols**, edited by *Amanda E. I. Proudfoot, Timothy N. C. Wells, and Christine Power, 2000*
137. **Developmental Biology Protocols, Volume III**, edited by *Rocky S. Tuan and Cecilia W. Lo, 2000*
136. **Developmental Biology Protocols, Volume II**, edited by *Rocky S. Tuan and Cecilia W. Lo, 2000*
135. **Developmental Biology Protocols, Volume I**, edited by *Rocky S. Tuan and Cecilia W. Lo, 2000*
134. **T Cell Protocols:** *Development and Activation*, edited by *Kelly P. Kearse, 2000*
133. **Gene Targeting Protocols**, edited by *Eric B. Kmiec, 1999*
132. **Bioinformatics Methods and Protocols**, edited by *Stephen Misener and Stephen A. Krawetz, 1999*
131. **Flavoprotein Protocols**, edited by *S. K. Chapman and G. A. Reid, 1999*
130. **Transcription Factor Protocols**, edited by *Martin J. Tymms, 1999*
129. **Integrin Protocols**, edited by *Anthony Howlett, 1999*
128. **NMDA Protocols**, edited by *Min Li, 1999*
127. **Molecular Methods in Developmental Biology:** *Xenopus and Zebrafish*, edited by *Matthew Guille, 1999*
126. **Adrenergic Rewceptor Protocols**, edited by *Curtis A. Machida, 2000*
125. **Glycoprotein Methods and Protocols:** *The Mucins*, edited by *Anthony P. Corfield, 2000*
124. **Protein Kinase Protocols**, edited by *Alastair Reith, 1999*
123. **In Situ Hybridization Protocols (2nd ed.)**, edited by *Ian A. Darby, 1999*
122. **Confocal Microscopy Methods and Protocols**, edited by *Stephen W. Paddock, 1999*
121. **Natural Killer Cell Protocols:** *Cellular and Molecular Methods*, edited by *Kerry S. Campbell and Marco Colonna, 1999*
120. **Eicosanoid Protocols**, edited by *Elias A. Lianos, 1999*
119. **Chromatin Protocols**, edited by *Peter B. Becker, 1999*
118. **RNA–Protein Interaction Protocols**, edited by *Susan R. Haynes, 1999*
117. **Electron Microscopy Methods and Protocols**, edited by *M. A. Nasser Hajibagheri, 1999*
116. **Protein Lipidation Protocols**, edited by *Michael H. Gelb, 1999*
115. **Immunocytochemical Methods and Protocols (2nd ed.)**, edited by *Lorette C. Javois, 1999*
114. **Calcium Signaling Protocols**, edited by *David G. Lambert, 1999*
113. **DNA Repair Protocols:** *Eukaryotic Systems*, edited by *Daryl S. Henderson, 1999*
112. **2-D Proteome Analysis Protocols**, edited by *Andrew J. Link 1999*
111. **Plant Cell Culture Protocols**, edited by *Robert D. Hall, 1999*
110. **Lipoprotein Protocols**, edited by *Jose M. Ordovas, 1998*
109. **Lipase and Phospholipase Protocols**, edited by *Mark H. Doolittle and Karen Reue, 1999*
108. **Free Radical and Antioxidant Protocols**, edited by *Donald Armstrong, 1998*
107. **Cytochrome P450 Protocols**, edited by *Ian R. Phillips and Elizabeth A. Shephard, 1998*
106. **Receptor Binding Techniques**, edited by *Mary Keen, 1999*
105. **Phospholipid Signaling Protocols**, edited by *Ian M. Bird, 1998*
104. **Mycoplasma Protocols**, edited by *Roger J. Miles and Robin A. J. Nicholas, 1998*
103. **Pichia Protocols**, edited by *David R. Higgins and James M. Cregg, 1998*
102. **Bioluminescence Methods and Protocols**, edited by *Robert A. LaRossa, 1998*
101. **Mycobacteria Protocols**, edited by *Tanya Parish and Neil G. Stoker, 1998*
100. **Nitric Oxide Protocols**, edited by *Michael A. Titheradge, 1998*
99. **Human Cytokines and Cytokine Receptors**, edited by *Reno Debets and Huub Savelkoul, 1999*
98. **Forensic DNA Profiling Protocols**, edited by *Patrick J. Lincoln and James M. Thomson, 1998*
97. **Molecular Embryology:** *Methods and Protocols*, edited by *Paul T. Sharpe and Ivor Mason, 1999*
96. **Adhesion Protein Protocols**, edited by *Elisabetta Dejana and Monica Corada, 1999*
95. **DNA Topoisomerases Protocols:** *II. Enzymology and Drugs*, edited by *Mary-Ann Bjornsti and Neil Osheroff, 1999*
94. **DNA Topoisomerases Protocols:** *I. DNA Topology and Enzymes*, edited by *Mary-Ann Bjornsti and Neil Osheroff, 1999*
93. **Protein Phosphatase Protocols**, edited by *John W. Ludlow, 1998*
92. **PCR in Bioanalysis**, edited by *Stephen J. Meltzer, 1998*
91. **Flow Cytometry Protocols**, edited by *Mark J. Jaroszeski, Richard Heller, and Richard Gilbert, 1998*
90. **Drug–DNA Interaction Protocols**, edited by *Keith R. Fox, 1998*
89. **Retinoid Protocols**, edited by *Christopher Redfern, 1998*
88. **Protein Targeting Protocols**, edited by *Roger A. Clegg, 1998*
87. **Combinatorial Peptide Library Protocols**, edited by *Shmuel Cabilly, 1998*
86. **RNA Isolation and Characterization Protocols**, edited by *Ralph Rapley and David L. Manning, 1998*
85. **Differential Display Methods and Protocols**, edited by *Peng Liang and Arthur B. Pardee, 1997*
84. **Transmembrane Signaling Protocols**, edited by *Dafna Bar-Sagi, 1998*
83. **Receptor Signal Transduction Protocols**, edited by *R. A. John Challiss, 1997*
82. **Arabidopsis Protocols**, edited by *José M Martinez-Zapater and Julio Salinas, 1998*

METHODS IN MOLECULAR BIOLOGY™

Molecular Methods in Developmental Biology

Xenopus *and Zebrafish*

Edited by

Matthew Guille

Institute of Biomolecular and Biomedical Sciences
University of Portsmouth
Portsmouth, UK

Humana Press ✹ **Totowa, New Jersey**

For
Kath, Connor, and Graihagh

© 1999 Humana Press Inc.
999 Riverview Drive, Suite 208
Totowa, New Jersey 07512

All rights reserved. No part of this book may be reproduced, stored in a retrieval system, or transmitted in any form or by any means, electronic, mechanical, photocopying, microfilming, recording, or otherwise without written permission from the Publisher. Methods in Molecular Biology™ is a trademark of The Humana Press Inc.

The content and opinions expressed in this book are the sole work of the authors and editors, who have warranted due diligence in the creation and issuance of their work. The publisher, editors, and authors are not responsible for errors or omissions or for any consequences arising from the information or opinions presented in this book and make no warranty, express or implied, with respect to its contents.

This publication is printed on acid-free paper. ∞
ANSI Z39.48-1984 (American Standards Institute) Permanence of Paper for Printed Library Materials.

Cover illustrations: Fig. 4E from "Cell and Tssue Transplantation in Zebrafish Embryos" by T. Mizuno, M. Shinya, and H. Takeda, and Fig. 1B from "Zebrafish Immunohistochemistry" by R, Macdonald.

Cover design by Patricia F. Cleary.

For additional copies, pricing for bulk purchases, and/or information about other Humana titles, contact Humana at the above address or at any of the following numbers: Tel: 973-256-1699; Fax: 973-256-8341; E-mail: humana@humanapr.com, or visit our Website at www.humanapress.com

Photocopy Authorization Policy:
Authorization to photocopy items for internal or personal use, or the internal or personal use of specific clients, is granted by Humana Press Inc., provided that the base fee of US $10.00 per copy, plus US $00.25 per page, is paid directly to the Copyright Clearance Center at 222 Rosewood Drive, Danvers, MA 01923. For those organizations that have been granted a photocopy license from the CCC, a separate system of payment has been arranged and is acceptable to Humana Press Inc. The fee code for users of the Transactional Reporting Service is: [0-89603-790-8/99 $10.00 + $00.25].

Printed in the United States of America. 10 9 8 7 6 5 4 3 2 1

Library of Congress Cataloging in Publication Data

Main entry under title: Methods in molecular biology™.

Molecular methods in developmental biology : Xenopus and zebrafish / edited by Matthew Guille.
        p.   cm.—(Methods in molecular biology ; 127)
    Includes index.
    ISBN 0-890603-790-8 (alk. paper)
    1. Embryology—Vertebrates—Laboratory manuals. 2. Molecular biology—Technique.
    3. Xenopus—Embryos. 4. Zebra danio—Embryos. I. Guille, Mathew. II. Series: Methods in molecular biology (Totowa, N.J.) ; v. 127.
    QL959.M668   19999
                                                                         98-50533
    571.8'616—dc21   CIP

Preface

The process whereby a single cell, the fertilized egg, develops into an adult has fascinated people for centuries. Great progress in understanding that process, however, has been made in the last two decades, when the techniques of molecular biology have become available to developmental biologists. By applying these techniques, the exact nature of many of the interactions responsible for forming the body pattern are now being revealed in detail. Such studies are a large, and it seems ever-expanding, part of most life-science groups. It is at newcomers to this field that this book is primarily aimed.

A number of different plants and animals serve as common model organisms for developmental studies. In *Molecular Methods in Developmental Biology:* Xenopus *and Zebrafish*, a range of the molecular methods applicable to two of these organisms are described, these are the South African clawed frog, *Xenopus laevis*, and the zebrafish, *Brachydanio rerio*. The embryos of both of these species develop rapidly and externally, making them particularly suited to investigations of early vertebrate development. However, both *Xenopus* and zebrafish have their own advantages and disadvantages. *Xenopus* have large, robust embryos that can be manipulated surgically with ease, but their pseudotetraploidy and long generation time make them unsuitable candidates for genetics. This disadvantage may soon be overcome by using the diploid *Xenopus tropicalis*, and early experiments are already underway. The transparent embryos of zebrafish render them well-suited for *in situ* hybridization and immunohistochemistry, and good for observing mutations in genetic screens. Zebrafish are, however, quite difficult to keep and somewhat prone to disease. Since both organisms are suited to similar studies, the techniques applied to them overlap considerably. A number of laboratories also use both models; thus this methods manual includes experimental approaches that can be applied to both species. Indeed, some of these approaches are identical for *Xenopus* and zebrafish, for example, those for analysis of steady-state mRNA levels by RT-PCR or RNase protection. However, when there are substantial differences, as in whole-mount *in situ* hybridization, two methods are described.

The first two chapters deal with tissue explant and transplant techniques. These have been included not only because they represent one of the major uses of these embryos, but also because they are mainly used in conjunction with the molecular techniques described here, for instance, in ectopic protein expression or RNA analysis. This combination of molecular techniques and embryo explants has proven extremely powerful, especially in the study of secreted signaling molecules with a role in development.

When one begins a developmental investigation, possibly with a novel or poorly defined protein to study, discovering the role of this gene and its product within an embryo is often the goal. Quantitative and spatial analysis of mRNA, the subject of the next four chapters (Chaps. 3–6), is often the starting point when attempting to investigate the developmental function of a newly discovered gene or protein. If these studies reveal that an mRNA is expressed in restricted regions of the embryo during the early stages of development, or at discrete embryonic or larval stages, this is often an indicator that the gene and its product will repay further analysis. Genes so expressed may well have a role in controlling development, and studying their regulation may add to our knowledge of how the pathways underlying embryo patterning are controlled. Although such mRNA analysis is useful, it is important to bear in mind that the expression of mRNA and protein do not necessarily correlate. There are many examples of translational control in developmental systems. Also, proteins that are expressed maternally may be present outside the cells that contain their cognate mRNA. There are a number of examples (both unreported and within the literature) where stable proteins, which are very widely expressed quite late in development as a result of maternal transcription, show very restricted mRNA expression if assayed by *in situ* hybridization. Ensuring that RNA analysis does not mislead in this way requires that protein expression is analyzed by immunohistochemistry, the subject of the two next chapters (Chaps. 7 and 8).

Analyzing mRNA and protein expression gives very limited information about the function of a given gene in the embryo. However, perturbing the level or activity of a test protein will in many cases produce data that define its role. One current weakness of both *Xenopus* and zebrafish as developmental models is that gene inactivation by homologous recombination, a technique that has lead to the elucidation of the function of many genes in mice, is not available. Although the ablation of a particular gene product cannot be achieved, the over- or ectopic expression of a protein is relatively straightforward in embryos of both the species described here. Thus the lack of the ability to make specific mutants has in part been overcome by the expression of ingenious variants of the endogenous proteins under study. These dominant interfering mutant proteins disrupt the activity of their endogenous counterparts, and effectively test gene inactivation. The techniques and experimental design underlying this type of approach are described in Chaps. 9–12.

All the techniques described up to that point in *Molecular Methods in Developmental Biology:* Xenopus *and Zebrafish* are useful for studying proteins, whatever their function. The final five chapters are more specialized (Chap. 12 contains both expression and promoter analysis methods). These deal with the analysis of transcription control in early development. Such analysis is particularly appropriate in *Xenopus* and zebrafish since, especially in the former, the large size and number of embryos available make viable the biochemical analysis of transcription factors and their interactions with DNA. Such analysis is extremely difficult at the early stages of development in other model organisms.

Though *Molecular Methods in Development Biology:* Xenopus *and Zebrafish* is primarily aimed at newcomers to the field, we have wherever possible obtained authors for each chapter who are still working at the bench and continually developing their techniques. Thus we hope that experienced workers may also find it useful. As with almost every methods manual it has been impossible to include all the techniques used to study *Xenopus* and zebrafish development. For example, zebrafish genetics are not touched upon here—there are many excellent treatments of positional cloning techniques available. Nonetheless the methods presented will provide a sound basis for the majority of studies. Other resources are available to both *Xenopus* and zebrafish developmental biologists on the internet; good starting points are the *Xenopus* molecular marker resource maintained by the Vize lab at the University of Texas (http://vize222.zo.utexas.edu/) or the zebrafish fishnet maintained by Institute of Neuroscience at the University of Oregon (http://zfish.uoregon.edu/).

Finally, many thanks to all the authors who managed to cope with yet another pressure in their hectic research careers, to the staff at Humana Press for their patience and help, and to my family for putting up with even more paper than usual strewn around the house. May all your experiments be successful!

Matthew Guille

Contents

Preface v
Contributors xi

1 The Animal Cap Assay
  Jeremy Green 1
2 Cell and Tissue Transplantation in Zebrafish Embryos
  Toshiro Mizuno, Minori Shinya, and Hiroyuki Takeda 15
3 Ribonuclease Protection Analysis of Gene Expression in *Xenopus*
  Craig S. Newman and Paul A. Krieg 29
4 Quantitative Analysis of mRNA Levels in *Xenopus* Embryos by Reverse Transcriptase–Polymerase Chain Reaction (RT-PCR)
  Oliver C. Steinbach and Ralph A. W. Rupp 41
5 Wholemount *In Situ* Hybridization of *Xenopus* and Zebrafish Embryos
  Joanne Broadbent and E. Mary Read 57
6 *In Situ* Hybridization to Sections of *Xenopus* Embryos
  David Bertwistle 69
7 Zebrafish Immunohistochemistry
  Rachel Macdonald 77
8 Immunohistochemistry of *Xenopus* Embryos
  Carl Robinson and Matthew Guille 89
9 Preparation and Testing of Synthetic mRNA for Microinjection
  Wendy Moore and Matthew Guille 99
10 Microinjection into *Xenopus* Oocytes and Embryos
   Matthew Guille 111
11 Microinjection into Zebrafish Embryos
   Qiling Xu 125
12 Expression from DNA Injected into *Xenopus* Embryos
   Ondine Cleaver and Paul A. Krieg 133
13 Promoter Analysis in Zebrafish Embryos
   Jos Joore 155

14 Transient Transgenesis in *Xenopus laevis* Facilitated by AAV-ITRs
   Yuchang Fu, Donghui Kan, and Sylvia Evans *167*
15 Band-Shift Analysis Using Crude Oocyte and Embryo Extracts from
   Xenopus laevis
   Rob Orford and Matthew Guille *175*
16 DNA Footprinting Using Crude Embryo Extracts from
   Xenopus laevis
   Rob Orford, Darren Mernagh, and Matthew Guille *187*
17 Mapping Protein–DNA Interactions Using In Vivo Footprinting
   David Warshawsky and Leo Miller *199*
Index *213*

Contributors

DAVID BERTWISTLE • *CRC Centre for Cell and Molecular Biology, The Institute of Cancer Research, UK*
JOANNE BROADBENT • *Developmental Biology Research Centre, King's College London, London, UK*
ONDINE CLEAVER • *Institute for Cellular and Molecular Biology and Zoology Department, University of Texas at Austin, Austin, TX*
SYLVIA EVANS • *Department of Medicine, University of California, San Diego, La Jolla, CA*
YUCHANG FU • *Department of Medicine, University of California, San Diego, La Jolla, CA*
JEREMY GREEN • *Division of Molecular Genetics, Dana-Farber Cancer Institute, Boston, MA*
MATTHEW GUILLE • *Division of Cell and Molecular Biology, Institute of Biomolecular and Biomedical Sciences, University of Portsmouth, Portsmouth, UK*
JOS JOORE • *Hubrecht Laboratory, Utrecht, The Netherlands (Present address: Westburg b.v., Leusden, The Netherlands)*
DONGHUI KAN • *Department of Medicine, University of California, San Diego, La Jolla, CA*
PAUL A. KRIEG • *Institute for Cellular and Molecular Biology and Zoology Department, University of Texas at Austin, Austin, TX*
RACHEL MACDONALD • *Department of Anatomy and Developmental Biology, University College London, London, UK (present address: SmithKline Beecham, Harlow, Essex, UK)*
DARREN MERNAGH • *Biophysics Laboratories, Institute of Biomolecular and Biomedical Sciences, University of Portsmouth, Portsmouth, UK*
LEO MILLER • *Department of Biological Sciences, University of Illinois, Chicago, IL*
TOSHIRO MIZUNO • *Division of Biological Science, Graduate School of Science, Nagoya University, Nagoya, Japan (Present address: Department of Chemistry and BioScience, Faculty of Science, Kagoshima University, Kagoshima, Japan)*

WENDY MOORE • *Biophysics Laboratories, Institute of Biomolecular and Biomedical Sciences, University of Portsmouth, Portsmouth, UK*
CRAIG S. NEWMAN • *Institute for Cellular and Molecular Biology and Zoology Department, University of Texas at Austin, Austin, TX*
ROB ORFORD • *Biophysics Laboratories, Institute of Biomolecular and Biomedical Sciences, University of Portsmouth, Portsmouth, UK*
E. MARY READ • *Developmental Biology Research Centre, King's College London, London, UK*
CARL ROBINSON • *Biophysics Laboratories, Institute of Biomolecular and Biomedical Sciences, University of Portsmouth, Portsmouth, UK*
RALPH A. W. RUPP • *Friedrich Mieschner Laboratorium der Max Planck-Gesellschaft, Tubingen, Germany*
MINORI SHINYA • *Division of Biological Science, Graduate School of Science, Nagoya University, Nagoya, Japan*
OLIVER C. STEINBACH • *Friedrich Mieschner Laboratorium der Max Planck-Gesellschaft, Tubingen, Germany*
HIROYUKI TAKEDA • *Division of Biological Science, Graduate School of Science, Nagoya University, Nagoya, Japan (Present address: Division of Early Embryogenesis, National Institute of Genetics, Mishima, Shizuoka, Japan)*
DAVID WARSHAWSKY • *Department of Biological Sciences, University of Illinois, Chicago, IL*
QILING XU • *Division of Developmental Neurobiology, National Institute for Medical Research, London, UK*

1

The Animal Cap Assay

Jeremy Green

1. Introduction

Over the last 10 years, the animal cap of the *Xenopus laevis* embryo has proved to be a versatile test tissue for a variety of molecules involved not only in animal development but also vertebrate cell regulation in general. These molecules include growth factors *(1–3)*, cell surface receptors *(4–6)*, signal transduction molecules *(7,8)*, transcription factors *(9)*, and extracellular matrix molecules *(10)*. The "animal cap assay" provides a simple, quick, inexpensive, and quantitative bioassay for biological activity of both cloned genes and purified or unpurified proteins.

The animal cap is a region of the *Xenopus* blastula and early gastrula stage embryo (6–12 h after fertilization). It is "animal" because the upper, pigmented half of the egg and embryo is referred to as the animal hemisphere (as opposed to the lower, vegetal hemisphere). The animal hemisphere is so named both because it contributes most to the final body (the vegetal hemisphere being mostly for yolk storage) and because those cells that it is made of are the most motile, or animated, during development. The animal cap is a "cap" because it forms the roof of a large cavity—the blastocoel—throughout blastula and gastrula stages. When excised and depending somewhat on the technique and stage of excision, it has the shape of a rather untidy skullcap.

The animal cap, if left *in situ*, normally contributes to the skin and nervous system of the tadpole. When excised and cultured in normal amphibian media (simple saline solutions), it develops into a ball of skin tissue or "atypical epidermis." The basis of the animal cap assay is that the excised animal cap can be diverted from its epidermal fate to other fates by (a) juxtaposition with other tissues, (b) inclusion of soluble growth factors or other reagents in the medium, or (c) by preinjecting the embryo with RNA or DNA encoding developmentally active genes. Importantly, the *Xenopus* animal cap does not respond promiscuously to nonspecific biological perturbation (*see* **Note 1**). Furthermore, it can respond in a number of informatively different ways to molecules that are active; for example, the response might be a change of cell type to neural, mesodermal, or endodermal fate. It might also include a morphological response, such as elongation. Another strength of the assay is that it can be made quantitative. Serial dilution of the test reagent and use of an objective scoring criterion (such as elongation) has proved very effective in quantitating amounts of active ingredient; for example, the mesoderm-inducing growth factor activin causes dramatic elongation of animal caps and is routinely quantitated by making a twofold dilution series and scoring (plus or minus) for any induction detectable as a morphological difference from uninduced control caps *(11,12)*.

Although the animal cap assay is a very useful one, some caution and a knowledge of the history of its use is advisable (*see* **Note 2**). The history begins with the discovery by Spemann in the 1920s that a transplanted amphibian dorsal lip, or Organizer, can induce a complete extra body axis in its host. The most prominent feature of the induced axis is an extra nervous system. In the 1930s, the hunt for the active ingredient in this inductive process ended in failure because the assay—essentially an animal cap assay—showed too many false-positive responses. This was because the experiments were done with newt and salamander embryos, not *Xenopus* embryos. In a number of amphibian species, the animal cap has a strong intrinsic tendency to become neuralized. Importantly, this is not the case for *Xenopus*. The *Xenopus* animal cap assay came to prominence when a number of laboratories were trying to identify the active molecule in the mesoderm induction. Nieuwkoop showed that whereas juxtaposition of an animal cap with Spemann's Organizer induces it to become neural tissue, juxtaposition of a cap with the vegetal hemisphere induces it to become mesoderm. Prominently induced among mesodermal tissues is skeletal muscle. In the mid-1980s, mesoderm induction was achieved with soluble growth factors, specifically fibroblast growth factor (FGF) *(13)* and what later turned out to be activin, a member of the transforming growth factor beta (TGFβ) superfamily of factors *(2,14)*. These two factors induce different spectra of mesodermal cell types and morphological responses. The dose (i.e., concentration and time of incubation) of growth factor is also critical in specifying the kind of response *(15)*. With the identification of mesoderm-inducing factors and the cloning of genes encoding them, it soon became routine to induce caps by injecting in vitro-transcribed RNA into embryos in the first few cell cycles and subsequently excising caps and incubating them without further additions.

The animal cap is not a uniform tissue, nor does its specification as epidermis represent an absolute cellular "default" or ground state. Its outer cells are different from its inner cells and its dorsal half is different from its ventral half by a number of criteria. Outer layer cells are pigmented, linked by tight junctions, and are relatively insensitive to mesoderm induction compared to the inner layer cells. Dorsal half-caps (as identified by labeling the embryo's and cap's dorsal side before explantation) are more readily induced to make dorsal mesoderm and neuroectoderm than the ventral half-caps. The difference is thought to be due to the epidermalizing effects of ventrally expressed bone morphogenetic protein 4 (BMP4) *(16–19)*. Cell dissociation (by incubation of animal caps in a medium lacking calcium) abolishes the dorsoventral differences, presumably by dispersing the secreted BMP.

The apparently complex biology of the animal cap response is an indication of how little is known about the ramified regulatory networks that are undoubtedly involved in the regulation of early development. The animal cap assay serves purely as a screen or assay for some biological activity—for example, in a screen or purification protocol for new genes and proteins—or as the focus in a study of early patterning of the ectoderm, mesoderm, and, even, endoderm.

2. Materials

1. A dissecting microscope (e.g., Nikon SMZ-U or a similar dissecting 10 W-power zoom microscope).
2. Cold light source (e.g., Schott KL1500 or similar fiber-optic "gooseneck" illuminator).
3. A controlled temperature (refrigerated) incubator (13–25°C).
4. A cooled dissection stage is helpful but not essential to prolong the period during which the embryos may be injected if microinjection is required.
5. In vitro fertilization with testis is normal to produce large numbers of synchronous embryos.
6. Dejellying of embryos is essential and carried out with 2% cysteine (pH 7.9–8.1 with sodium hydroxide). Dejellying after two or three cell divisions is recommended, because it is then easy and desirable to remove sick embryos and unfertilized eggs and to keep the good embryos well dispersed to maximize synchrony.
7. 1X Marc's Modified Ringers (MMR): 100 mM NaCl, 2 mM KCl, 2 mM CaCl$_2$, 1 mM MgCl$_2$, 10 mM HEPES pH 7.4 (*see* **Note 3**).
8. Plastic Petri dishes lined with fresh 1% agarose (*see* **Note 4**).
9. Fine watchmaker's forceps, such as Dumont number 5 "Biologie" forceps, are essential for removal of the outer "vitelline" membrane of the embryo and for excision of the cap. (Tungsten or glass needles can also be used, but the dissection is slower and not significantly more precise than using forceps.). The forceps can be used "straight out of the box," but a little sharpening on a piece of wet–dry abrasive paper or a sharpening stone is helpful in improving or restoring the forceps tips. Note, however, that the sharpening should be minimal (perhaps

two or three gentle strokes of the tips angled at about 30° to the horizontal surface) and done with the forceps tips held together to maintain the meeting points.
10. Pipets: the ends are broken off Pasteur pipets (after scoring with a diamond pencil) to leave a mouth 3–4 mm in diameter. For moving explants, an unmodified Pasteur pipet can be used, although a Gilson Pipetman P10 with a cut off yellow tip is also suitable and somewhat easier to control. For removing explants from the rather deep wells of a multiwell plate, it is a good idea to use a Pasteur pipet that has been bent over a flame.

3. Methods
3.1. Test Material

1. For soluble proteins or protein mixtures, serial twofold dilutions should be prepared in the 1X MMR, 0.1% bovine serum albumin (BSA). If the test substance is prepared in its own medium (e.g., conditioned tissue culture medium, then care must be taken that this medium does not significantly alter the composition of the MMR. Thus, either use dilutions of greater than 1 in 10, dialyze the test substance, or use ultrafiltration and dilution before adding it to MMR.
2. For RNA injections, the RNA is transcribed from a suitable linearized DNA template using an in vitro transcription kit (Message Machine, Ambion, Austin, TX) or components bought separately (*see* **ref. 20**, Chapter 9). RNA is phenol extracted and ethanol precipitated and quantified carefully. We usually quantify RNA on an ethidium–agarose electrophoresis gel against spectrophotometrically quantified RNA standards. This gives information about integrity as well as quantity. RNAs are injected in amounts varying from 5 pg to 5 ng per embryo to obtain biological effects. It is important to include water-injected and nonsense RNA controls to check for nonspecific effects of the injection. It is very important to note that RNA injected in the one- to two-cell stage embryo and later does not diffuse freely from the site of injection, so that for animal cap assays, the RNA must be injected in the animal hemisphere.

3.2. Embryo Preparation and Explantation

The animal cap excision day falls into one of two patterns. Either eggs are fertilized in the evening and kept at 13–14°C overnight for dissection the following morning, or they are fertilized in the early morning and kept at room temperature or warmer (up to 25°C) for dissection the same day. The evening fertilization is recommended for analysis at gastrula stages, as these are reached in the afternoon or evening of the dissection day. The number of caps to be excised must be estimated together with the stage at which they will be dissected (*see* **Notes 5** and **6**).

1. Embryos must be well dejellied to enable removal of the vitelline membrane. About 6 min at room temperature in 2% cysteine pH 8.0 is typically sufficient to do this.

2. The removal of the vitelline membrane or envelope is the hardest step in the animal cap assay. The following steps provide a description of one approach, but such a description in words is inevitably a poor substitute for laboratory demonstration by an expert (*see* **Fig. 1**). Lots of practice is essential in any case to develop a "feel" for the procedure. Be warned that the novice will inevitably mash the first few dozens of embryos before a single clean "devitellinization" is successfully achieved. Fortunately, for an animal cap assay it does not matter if the entire vegetal and marginal regions of the embryo are obliterated as long as the cap itself is intact. Set up the lighting under the dissection microscope to show of the brilliant shine or glint at the embryo surface. This bubblelike shine is due to the vitelline membrane. The membrane itself is quite hard to see, and the glint of reflected light is very helpful in tracking it.
3. Grasp the membrane with the very tips of one pair of forceps in the marginal or vegetal region while bracing the embryo against the side of the other forceps. The vitelline membrane is slippery and the embryo has a tendency to roll with vegetal pole down. Thus, the grabbing/bracing movement has to be coordinated and quite quick. Ideally, the membrane is grabbed cleanly without penetrating the embryo itself, but almost inevitably one of the forceps tips stabs through the membrane and into the yolky vegetal cells. This does not matter as long as a firm grasp of the membrane is achieved.
4. With the other forceps, grasp for the membrane close to where the first pair penetrates and holds the membrane and pull away from the first with a looping movement. This second grasp is best done essentially "blind," in that the optimal grabbing point is invisible but always at the surface of the first forceps, just behind the first forceps' tips. The looping movement should trace the curvature of the embryo surface at about one embryo diameter's distance from it. The best direction for the looping action will vary from embryo to embryo. This action and distance tears the membrane and maximizes the length of the tear without ripping the embryo itself. Repeating **step 3** may be necessary, but with one or two such rips, the vitelline membrane should be loosened and crumpled such that is easy to grab and pull off the embryo with either of the forceps.
5. After vitelline membrane removal, it is a good idea to roll the embryo animal pole up and gently push it back into shape. This helps maintain a good blastocoel, which eases cap explantation. It also prevents contact between the animal cap and the blastocoel floor, which can lead to mesoderm induction.
6. Before excising the cap, it is important to estimate the location of the animal pole and blastocoel. Gently prod the devitellinized embryo to reveal where the blastocoel is, because overlying pigmented tissue is more easily depressible than neighboring marginal regions. Care must be taken to take only animal cap tissue and not marginal zone material because the latter is specified very early in development to become mesoderm. Marginal zone cells are easily recognized because they are larger and more yolky that animal cap cells. If accidentally excised with the animal cap, they should be trimmed off.
7. Make V-shaped cuts around the animal pole using forceps. The cuts are made by pinching the devitellinized embryo about halfway between animal pole and equa-

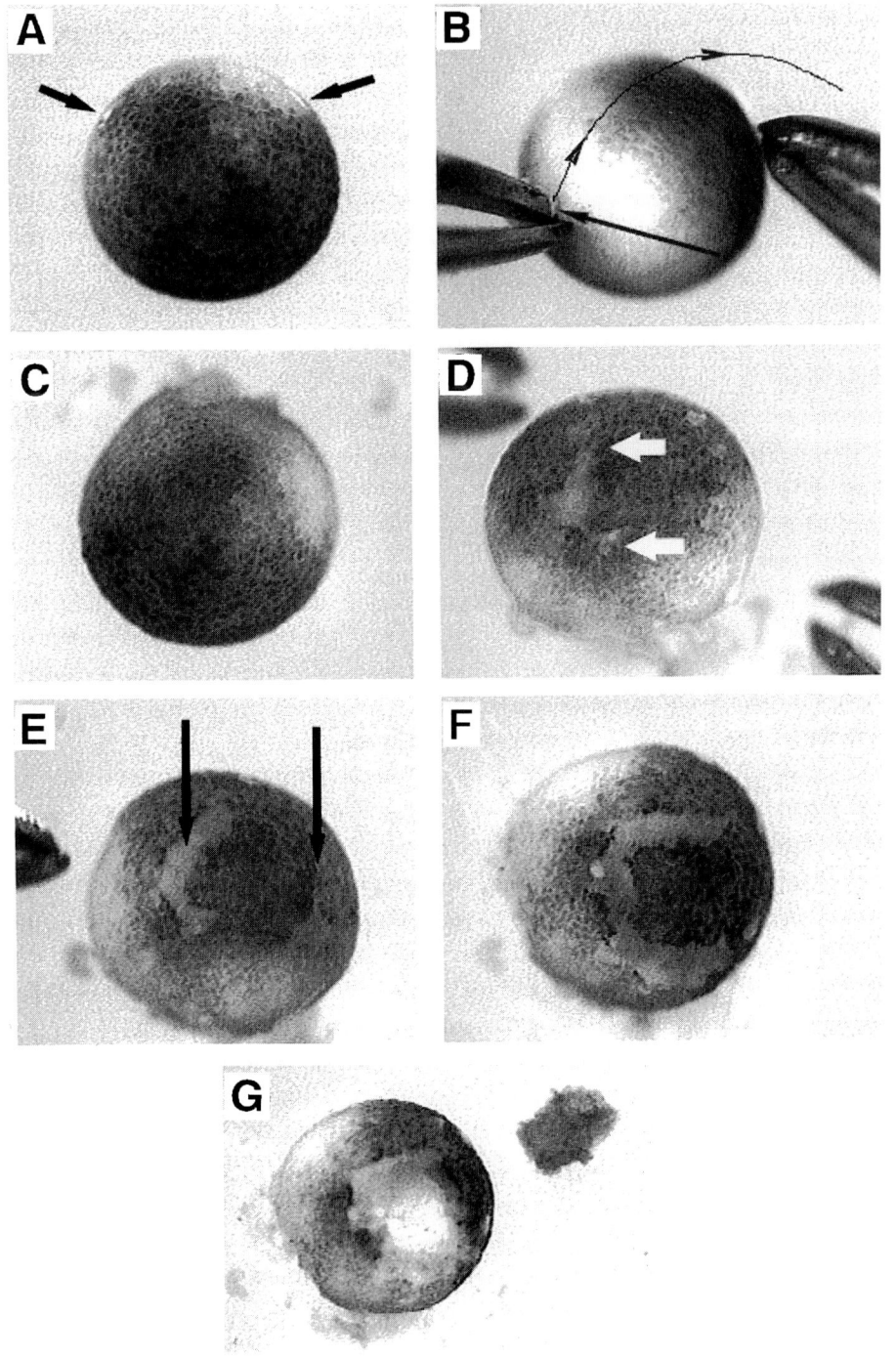

tor. A darting movement made during the pinching action gives a cleaner cut and prevents sticking of tissue to the forceps. Make a cut first with one pair of forceps, then at a diametrically opposite position with the other. Rotate the embryo 90° clockwise or anticlockwise and make two more cuts. The cap should lift out from the embryo with the last pinching movement. With practice, the forceps pinching method can be as neat and easy as most of the alternative dissection methods (*see* **Note 7**) and is certainly much faster.

8. It is important for induction by soluble factors to transfer animal caps to the inducer-containing medium soon (i.e., within a few minutes) after excision. As soon as caps are excised in calcium-containing medium, they begin to curl up at the edges. Eventually, they roll up into a ball that is impervious to induction by growth factors subsequently added to the medium *(11)*. This "rounding up" is faster in some embryo batches than others, but typically takes place over 10–20 min. The rounding up may be delayed in low-calcium medium, but this is not recommended because once a cap starts to round up, it goes to completion quite quickly regardless of the medium.

9. Incubation time depends on what is to be assayed. It is critical that sibling whole embryos are kept at the same temperature to monitor developmental stage. Caps seem to do best when incubated at 18°C, slightly cooler than room temperature. However, this is not a strong effect and the temperature should be adjusted to facilitate harvesting at the appropriate stage.

10. Harvest the explants at the appropriate stage below (*see* **Note 8**).

Fig. 1. Steps in animal cap excision using the two-forceps method. (**A**) A stage 8.5 blastula. Note the shining highlights on the vitelline membrane (arrows). (**B**) The embryo is braced with the right forceps while the vitelline membrane is grabbed by the left forceps. The upper point of the left forceps has penetrated the membrane (tip of straight arrow). The right forceps are brought to grasp at the vitelline membrane just where the left forceps penetrate or meet the embryo surface. Upon grasping, the right forceps are drawn upward and to the right (curved line) in a looping motion. (**C**) The devitellinized blastula is rolled and shaped so that its animal pole is once again uppermost and it is nearly spherical. Note differences between this and the blastula in panel **A,** namely no glinting membrane and a flatter, more spread out shape. Debris has leaked from the vegetal pole and is lying around the embryo, but it does not affect the animal cap. (**D**) After the first pinching cut with the left forceps. White arrows mark where the forceps points first penetrated the animal hemisphere and the limits of the "<"-shaped cut. (**E**) After the second cut using the right forceps. The right incision is hard to see in this example, but note that the distance between the cuts encompasses only the middle 50% of the embryo diameter. (**F**) After rotating the embryo clockwise 90°, a third cut (using the left forceps) produces the "trapdoor" appearance. (**G**) The pinching action of the fourth cut pulls out the animal cap, on the right. Note the relatively dark color of the inner surface of the animal cap (showing) compared to the very light, yolky blastocoel floor.

Stage	Assay	Purpose
10.5	RNA	Transcription of "immediate early" genes
12–18	RNA, immunostaining	Analysis of early patterning (e.g., Hox) genes
13–15	Inspection	Elongation (transient for FGF, sustained for activin)
25 onward	RNA, immunostaining, histology	Terminal differentiation
25 onward	Visual inspection	Elongation or "balloon" formation

4. Notes

1. There is a philosophical objection to the animal cap assay, namely that because the animal cap's normal specification is to become epidermal, any change to this is somehow nonphysiological. This argument is, of course, undeniable, but it is not an objection to the animal cap assay as such. Instead, it is an important fact to be borne in mind when choosing among alternative assays and in interpreting data that the animal cap assay generates. Some of the past discoveries about the animal cap (*see* **Subheading 1**) have shown that it is not a homogenous "naive" tissue nor a static one. Some of its salient features are worth reiterating:
   a. Dorsoventral asymmetry (the dorsal half of an animal cap is much easier to induce to make, for example, dorsal mesoderm than the ventral half)
   b. Inside–outside asymmetry (outer, pigmented cap cells are less responsive to some mesoderm inducers than inner blastocoel roof cells, whereas outer cells may be more responsive to other types of induction such as cement gland)
   c. Transient sensitivities (responsiveness to mesoderm inducers declines gradually during the beginning of gastrulation; responsiveness to Xwnt8 expression seems to change as early as the midblastula stage)
   To these should be added some other less obvious properties:
   d. Changing cell population (the cell movements known as epiboly mean that cells are constantly moving out of the animal cap into the marginal zone and thinning the cap itself)
   e. Changing extracellular matrix (by very late blastula and early gastrula stages, the cap becomes sticky to dissect, presumably because of deposition of fibronectin and other extracellular matrix components)

   Fortunately, it is relatively straightforward to control for these factors. Dorsoventral asymmetry can be abolished by ultraviolet-ventralising the embryos (*see* Chap. 14 of **ref. 20**). Inside–outside differences can be monitored histologically or made physically separate by cell dissociation. Timing factors can, and should, be investigated by taking caps at specific stages. As cap cutting itself can be quite quick, the time resolution of such experiments is good.

2. When should the cap assay be used? Very often, overexpression of a gene in a whole embryo leads to a complex and uninterpretable effect. The animal cap assay can often provide a simpler phenotype. This is particularly true if the ques-

tion being asked concerns direct and immediate effects of gene expression or protein application. Furthermore, this kind of "direct action" assay is much easier to do in *Xenopus* than in almost any other model embryo species.

Another type of use for the animal cap assay is as a pure assay, screen, or reporter without specific reference to normal cap physiology; for example, it can be used in tracing very low quantities of active proteins from non-*Xenopus* species during purification procedures. This type of use has not been greatly exploited because most *Xenopus* scientists are interested in the biology of the cap and factors themselves. Such a use depends, of course, on the material to be tested having some activity. However, the extreme sensitivity and speed of the assay should recommend it to a wider audience for such materials. Dissociating the cells of excised animal caps has been used extensively to control or eliminate cell–cell signaling and increase exposure of cells to soluble factors. For a detailed protocol, *see* **ref. 21**). Cells kept dissociated do not survive well and tend to differentiate as neural cells. Relatively transient dissociation maintains the epidermal specification of the cap while allowing other manipulations.

Caps can be used in screens for cloning. cDNA libraries are made in vectors that enable transcription of mRNA in vitro. The libraries are divided into pools (small pools of about 100 clones appear to be optimal). The pools are transcribed and the mRNA generated is injected into embryo or oocyte animal hemispheres. From embryos, the caps can be excised and simply assayed. For a paracrine screen (i.e., for secreted factors), a normal animal cap is placed hatlike onto the top of an injected oocyte. Such screens have been used successfully to identify and isolate genes of significant biological interest *(22)*.

Caps have been used to investigate the penetration of signals through tissue. One or more caps are juxtaposed with a known source of mesoderm-inducing signal. By lineage labeling either the responding cap or the signal source tissue (which can also be an injected cap) signal penetration or transmission through several cell diameters has been demonstrated *(23,24)*.

Caps have also been used to assay signals from chicken embryos. Caps wrapped in the chick's Hensen's node, for example, become neuralized. This assay has the advantage that the conjugated tissues are incubated at a little below room temperature, effectively freezing the chick's development while allowing the *Xenopus* tissue to develop and respond to chick signals *(25)*.

3. Any full-strength amphibian saline (e.g., MMR, normal amphibian medium [NAM]; *see* **ref. 20**) may be used. The high salt levels in these media cause whole embryos to exogastrulate, but in animal cap explants, they encourage healing. Other media can be used to delay "rounding up" of the explanted cap. This can be helpful experimentally, as rounding up can be rapid and fully rounded cap explants are not responsive to subsequently applied soluble factors. To prolong the process, a one-tenth dilution of MMR in calcium-magnesium-free medium (CMFM) is recommended *(20)*. However, it is extremely difficult to stop rounding up entirely and the rate of rounding varies from egg batch to egg batch. (If more controlled cell exposure is important, then a dissociated cell protocol is recommended.) If soluble pro-

tein factors are to be used in the medium, bovine serum albumin (BSA, Sigma) should be added to 0.1% w/v to block nonspecific protein binding.
4. The agarose lining of dissection and incubation plates prevents sticking of explants. Depending on the number of caps to be assayed, it is essential to have sufficient numbers of dissection dishes, as they quickly become full of yolky debris during dissection. At least one 35-mm dish per 20 embryos/caps to be dissected is recommended.

   Depending on the number of conditions and caps to be assayed, agarose-lined dishes or multiwell plates must also be prepared for the caps after dissection. A critical factor is that explants tend to fuse with one another, which can obscure observation of morphological responses. Cap fusing has two effects. One is that, like rounding up, it excludes penetration or access of soluble factors. The other is that scoring morphological changes is much harder in fused caps than in single caps. Where neither morphology nor factors in the medium are important, cap fusion seems to have little effect on, for example, gene expression. To keep explants separate, they can be assayed as one explant per well in an agarose-lined 96-well tissue culture plate. Alternatively, two or more explants can be placed in separate depressions made in agarose-lined dishes or larger wells. Depressions are made using the sealed, red-hot end of a glass Pasteur pipet or metal fork. Alternatively, they can be cast into the agarose as follows. A mold is drilled in a block of Teflon or similar material consisting of an array of 1.8-mm-diameter × 1.0-mm-deep depressions in the floor of a 4-mm-deep recess. The recess is slightly smaller that the Petri dish to be used for the embryos. A nonadhesive silicone compound, such as Dow-Corning Sylgard 184, is cast in the mould to generate a disk or square of rubber about 2 mm thick with 1-mm pimples on the underside. This is floated on the surface of molten 1% agarose and removed after the agarose has set, leaving depressions suitable for embryos and explants.
5. For straightforward morphological assays, such as elongation in response to activin, as few as two caps per condition is sufficient and gives reproducible and quantitative results. For some morphological assays, such as for FGF, several caps are required because the morphological response is weaker and more unreliable. For RNA analysis by reverse transcriptase–polymerase chain reaction (RT-PCR), one or two caps per condition is minimally sufficient. However, more caps will improve RNA yield per cap and enable duplicate assays for multiple genes—strongly recommended for RT-PCR. For RNA analysis by RNase protection assay (RPA), 10 caps per condition is advised. Although this seems like more work, RPA enables several genes to be quantitatively analyzed in the same tube. This provides better quantitative control than with RT-PCR. For wholemount *in situ* hybridization, the number of caps needed is largely a matter of taste, provided the gene expression is patently reproducible. Similarly, caps to be harvested for immunostaining or conventional histological staining should be numerous enough to allow for some losses during workup and for persuasive reproducibility to be apparent. Generally speaking, it is better to cut additional caps than to economize. With practice, it should be possible for an average worker to dissect 60–100 caps per hour.

6. A range of dissection stages is available. It is extremely difficult to dissect an animal cap before Nieuwkoop and Faber (NF) stage 6.5 because, until then, the blastocoel is very small and the animal cap consists of very few, large, fragile cells. Even at stage 7.0, results are less likely to be consistent than at stages 7.5–9.5. The response to soluble mesoderm-inducing factors is constant during the 7.5–9.5 window. After this time, with the onset of gastrulation (stage 10 onward), responsiveness to mesoderm inducers activin and FGF declines. Explantation is further complicated by the involution of mesoderm into the blastocoel underlying the animal cap. Animal cap that is underlain by mesoderm is respecified from epidermal to neural fate so that, although explantation is still possible, the nature of the explant and thus the assay changes. Toward stage 10, the animal cap also becomes sticky, and sticks to the forceps during dissection. Thus, the 3- to 4-h window between stages 7.5 and 9.5 (mid to late blastula) is both the most well-defined and the most convenient dissection period. If assays from caps dissected throughout this period are inconsistent, then more restricted ranges within this range should be compared.
7. There are two variations on the excision method described. One is to use different tools to make the same cuts; for example, instead of forceps, a sharpened tungsten needle can be used to make the cuts. The needle is inserted into the blastocoel and used to cut through it by pressing it up against the underside of either forceps or a second needle held at the cap surface. This method is slower than the forceps-only method and perhaps, because of this, can lead to neater cutting. However, when both methods are mastered, the difference is negligible. The second variation on the above excision method is more radical: The cap is cut from below after first inverting the embryo and then cutting open the blastocoel via vegetal hemisphere. The main merit of this approach is that the precise position of the blastocoel, cap, and marginal zone are apparent before the cap itself is excised. This prevents inclusion of any marginal zone cells in the explant. However, the method is very much slower and messier.

    Cap size and site of excision can be important for one main reason. Very large or off-center caps inevitably contain some marginal zone cells and can, in some circumstances, be more sensitive to induction than smaller caps. Thus, in general, it is better to err on the small side. However, caps can be too small. Very small caps are physically less robust and can fail to undergo morphological changes such as extension movements. Care is therefore required to make caps by cuts at a latitude of about 45° from the animal pole and thus about 0.5 mm across. Sizing the caps by eye (rather than, say, using a micrometer) is sufficient to get consistent results, although if this turns out to be a problem, one of the alternative excision methods might be appropriate. In any case, it is always a good idea to cut at least two caps for each condition to be assayed. The stage of excision also plays a role. The blastocoel is much larger in late blastula than early blastula and is thus easier to dissect cleanly.
8. For analysis of gene expression, it is important to know what the normal in vivo expression of a gene is before using it as part of an animal cap assay. The dynamic

nature of much gene expression means that the same gene in an animal cap can mean different things at different stages. If possible, more than one gene should be analyzed. Functional tests and differentiation itself must ultimately be more persuasive if the interpretation of gene expression is at all ambiguous. Expression of too few genes in animal caps is, if anything, overused and overinterpreted in the literature.

9. Animal caps can be embedded in wax and sectioned using standard procedures. The sectioning is somewhat difficult due to the small size of the samples. Thus, it is often preferable to do wholemount staining. Staining of these hard-to-handle explants is best done in small "baskets." These can be made by heat sealing 70-µm nylon or polyester mesh onto the end of a cut microfuge tube or both ends of a short section of Tygon tubing. Heat sealing is done on a piece of aluminum foil covering a hotplate. Rather large baskets called Netwell inserts (Costar) can also be used, although these require larger volumes of probe and antibody solution.

References

1. Kimelman, D. and Kirschner, M. (1987) Synergistic induction of mesoderm by FGF and TGF-beta and the identification of an mRNA coding for FGF in the early *Xenopus* embryo. *Cell* **51,** 869–877.
2. Smith, J. C., Price, B. M., Van Nimmen, K., and Huylebroeck, D. (1990) Identification of a potent *Xenopus* mesoderm-inducing factor as a homologue of activin A. *Nature* **345,** 729–731.
3. Hogan, B. L., Blessing, M., Winnier, G. E., Suzuki, N., and Jones, C. M. (1994) Growth factors in development: the role of TGF-beta related polypeptide signalling molecules in embryogenesis. *Development* **120,** 53–60.
4. Amaya, E., Musci, T. J., and Kirschner, M. W. (1991) Expression of a dominant negative mutant of the FGF receptor disrupts mesoderm formation in *Xenopus* embryos. *Cell* **66,** 257–270.
5. Hemmati-Brivanlou, A. and Melton, D. A. (1992) A truncated activin receptor inhibits mesoderm induction and formation of axial structures in *Xenopus* embryos. *Nature* **359,** 609–614.
6. Howard, J. E., Hirst, E. M., and Smith, J. C. (1992) Are beta 1 integrins involved in *Xenopus* gastrulation? *Mech. Devices* **38,** 109–119.
7. Graff, J. M., Bansal, A., and Melton, D. A. (1996) *Xenopus* Mad proteins transduce distinct subsets of signals for the TGF beta superfamily. *Cell* **85,** 479–487.
8. Massague, J. (1996) TGFbeta signaling: receptors, transducers, and Mad proteins. *Cell* **85,** 947–950.
9. Cunliffe, V. and Smith, J. C. (1992) Ectopic mesoderm formation in *Xenopus* embryos caused by widespread expression of a Brachyury homologue. *Nature* **358,** 427–430.
10. Brickman, MC and Gerhart, JC (1994) Heparitinase inhibition of mesoderm induction and gastrulation in Xenopus laevis embryos. *Dev. Biol.* **164,** 484–501.
11. Cooke, J., Smith, J. C., Smith, E. J., and Yaqoob, M. (1987) The organization of mesodermal pattern in *Xenopus laevis*: experiments using a Xenopus mesoderm-inducing factor. *Development* **101,** 893–908.

12. Smith, J. C., Yaqoob, M., and Symes, K. (1988) Purification, partial characterization and biological effects of the XTC mesoderm-inducing factor. *Development* **103,** 591–600.
13. Slack, J. M., Darlington, B. G., Heath, J. K., and Godsave, S. F. (1987) Mesoderm induction in early *Xenopus* embryos by heparin-binding growth factors. *Nature* **326,** 197–200.
14. Smith, J. C. (1987) A mesoderm inducing factor is produced by a *Xenopus* cell line. *Development* **99,** 3–14.
15. Green, J. B. A., Howes, G., Symes, K., Cooke, J., and Smith, J. C. (1990) The biological effects of XTC-MIF: quantitative comparison with *Xenopus* bFGF. *Development* **108,** 173–183.
16. Dale, L., Howes, G., Price, B. M., and Smith, J. C. (1992) Bone morphogenetic protein **4,** a ventralizing factor in early *Xenopus* development. *Development* **115,** 573–585.
17. Graff, J. M., Thies, R. S., Song, J. J., Celeste, A. J., and Melton, D. A. (1994) Studies with a *Xenopus* BMP receptor suggest that ventral mesoderm-inducing signals override dorsal signals in vivo. *Cell* **79,** 169–179.
18. Suzuki, A., Thies, R. S., Yamaji, N., Song, J. J., Wozney, J. M., Murakami, K., and Ueno, N. (1994) A truncated bone morphogenetic protein receptor affects dorsal-ventral patterning in the early *Xenopus* embryo. *Proc. Natl. Acad. Sci. USA* **91,** 10,255–10,259.
19. Maeno, M., Ong, R. C., Suzuki, A., Ueno, N., and Kung, H. F. (1994) A truncated bone morphogenetic protein 4 receptor alters the fate of ventral mesoderm to dorsal mesoderm: roles of animal pole tissue in the development of ventral mesoderm. *Proc. Natl. Acad. Sci. USA* **91,** 10,260–10,264.
20. Peng, H. B. and Kay, B. K. (eds.) (1991) *Xenopus laevis*: practical uses in cell and molecular biology, in *Methods in Cell Biology*, Academic, New York.
21. Green, J. B. A., New, H. V., and Smith, J. C. (1992) Responses of embryonic *Xenopus* cells to activin and FGF are separated by multiple dose thresholds and correspond to distinct axes of the mesoderm. *Cell* **71,** 731–739.
22. Lustig, K. D., Kroll, K. L., Sun, E. E., and Kirschner, M. W. (1996) Expression cloning of a *Xenopus* T-related gene (Xombi) involved in mesodermal patterning and blastopore lip formation. *Development* **122,** 4001–4012.
23. Gurdon, J. B., Harger, P., Mitchell, A., and Lemaire, P. (1994) Activin signalling and response to a morphogen gradient. *Nature* **371,** 487–492.
24. Gurdon, J. B., Mitchell, A., and Mahony, D. (1995) Direct and continuous assessment by cells of their position in a morphogen gradient. *Nature* **376,** 520–521.
25. Kintner, C. R. and Dodd, J. (1991) Hensen's node induces neural tissue in *Xenopus* ectoderm. Implications for the action of the organizer in neural induction. *Development* **113,** 1495–1505.

2

Cell and Tissue Transplantation in Zebrafish Embryos

Toshiro Mizuno, Minori Shinya, and Hiroyuki Takeda

1. Introduction

Zebrafish (*Danio rerio*) embryos have gained considerable popularity in recent years because they offer several advantages for developmental studies. The embryos are easy to manipulate, develop quite rapidly, and many genetic mutations are now becoming available. Classical cell and tissue transplantation techniques have been frequently applied to zebrafish embryos to analyze the state of cell commitment, inductive interaction between embryonic tissues and defective tissues in mutant embryos. This chapter introduces three kinds of transplantation techniques useful for the analysis of early inductive events in zebrafish embryos, such as mesoderm and neural induction.

In the first, the technique for yolk cell transplantation is described. In the teleost embryo, a large yolk cell is located vegetally, under the blastoderm which forms the embryo proper. It has been suggested that substances are passed from the yolk cell to the blastoderm to induce embryonic axes *(1)*. To examine the inductive properties of the yolk cell, we have developed a transplantation method. By use of this technique, it has been demonstrated that, as in amphibian vegetal cells, the yolk cell of the teleost is responsible for induction and dorsoventral patterning of the mesoderm *(2)*. Thus, normal activity of the yolk cell is essential for the early development of zebrafish. The technique will be useful in analyzing mutants showing defects in the embryonic axes, as the inductive activity of the yolk cell could be affected in some of those mutants.

The second technique has been developed in order to produce ventralized fish embryos. Ventralized embryos, in which maternal dorsal determinants have been inactivated or removed, have been an effective tool for analyzing

From: *Methods in Molecular Biology, Vol. 127: Molecular Methods in Developmental Biology: Xenopus and Zebrafish* Edited by: M. Guille © Humana Press Inc., Totowa, NJ

the mechanism underlying dorsoventral axis formation. In *Xenopus*, the embryos resulting from ultraviolet (UV) irradiation to the vegetal hemisphere of fertilized eggs show a ventralized phenotype, in which little or no axial structures are formed *(3)*. By contrast, UV irradiation also causes incomplete epiboly in zebrafish embryos *(4)*. Thus, until recently, no reliable method of producing ventralized embryos was available in zebrafish. We found, however, that ventralized fish embryos were reproducibly obtained by the removal of the vegetal yolk cell mass soon after fertilization. This method was developed based on the fact that teleost cytoplasmic determinants involved in induction of dorsal tissues are localized at the vegetal end of the yolk cell at the time of fertilization *(5)*. They are then translocated from the vegetal end to the future dorsal side under the blastoderm during cleavage stages. This movement of the determinants is reminiscent of cortical rotation in amphibian embryos which occurs soon after fertilization and is blocked by UV irradiation *(6)*. This technique assures a complete removal of dorsal determinants and can be used to analyze dorsoventral patterning in the fish embryo.

Finally, we describe a tissue transplantation technique similar to that described elsewhere *(7)*. We, therefore, focus on the transplantation of organizer tissues which can be used for the analysis of neural induction in zebrafish. Furthermore, we found that, when transplanted into zebrafish embryos, mammalian cultured cells producing organizer factors mimicked the endogenous organizer. The transplantation of cultured cells is widely applicable. If a gene of interest encodes a secreted factor, its role in vivo can be easily assessed by transplanting cultured cells which have been transfected with the appropriately expressing cDNA into embryos.

2. Materials

1. Micropipet: The glass capillaries (blunt end tip, $\varnothing = 1$ mm (e.g., Narishige [Tokyo, Japan], G-1) are pulled to fine tips on a electrode puller (e.g., Narishige, PN-3). The tips are broken off at an angle using a hand-held razor blade. Capillary glass which contains an internal filament cannot be used because the filament may destroy cells during the transplantation procedure. The tips can be fire polished with a microforge (e.g., Narishige, MF-9), or the micropipet can be used without fire polishing the tip. The diameter of the tip for shield transplantation is 30–50 µm.
2. Phosphate-buffered saline (PBS): 8 g NaCl, 0.2 g KCl, 2.9 g $Na_2HPO_4 \cdot 12H_2O$, 0.2 g KH_2PO_4 in 1 L (pH 7.2).
3. 1X Ringer's solution: 116 m*M* NaCl, 2.9 m*M* KCl, 1.8 m*M* $CaCl_2$, 5 m*M* HEPES (pH 7.2).
4. (1/3)X Ringer's solution: 39 m*M* NaCl, 0.97 m*M* KCl, 0.6 m*M* $CaCl_2$, 1.67 m*M* HEPES (pH 7.2).
5. Calcium-free (1/3)X Ringer's solution: 39 m*M* NaCl, 0.97 m*M* KCl, 10 m*M* EDTA, 1.67 m*M* HEPES (pH 7.2).

6. Agar (e.g., DIFCO [Frankiln Lakes, NJ] BACTOAGAR): dissolved in distilled water or the desired Ringer's solution.
7. Antibiotics: penicillin and streptomycin solution (10000 U/mL penicillin and 10,000 µg/mL streptomycin, Gibco BRL [Rockville, MD] 15140-122) are added to all media used for operations at a final concentration of 1% to 2%.
8. Methyl cellulose (e.g., 3500–5600 cps, Sigma [St. Louis, MO] M-0387).
9. Lipofectamine™ (Gibco BRL 18324-012).
10. Rhodamin-dextran (10,000 MW, e.g., Molecular Probes, [Eugene, OR] D-1816).
11. Biotin-dextran (10,000 MW, lysine fixable; e.g., Molecular Probes, D-1956).
12. Albumen, prepared from egg white: Addition of egg albumen to Ringer's solution sometimes increases the survival rate of embryos which have been manipulated, especially when the embryos have sustained some damage to the yolk membrane by the removal of the yolk or fusion of two embryos *(8)*. In addition to nutritive components, the albumen contains *lysozyme*, a bacteriostatic agent. For this reason, egg albumen is often used in embryo cultures to prevent the growth of microorganisms as well as for nutrition.
13. Embryo transfer pipet: Pasteur pipets and rubber teats.
14. 35-mm, 60-mm, and 100-mm plastic culture dishes with lids.
15. Agar-coated dishes for dechorinated embryos: Pour an appropriate amount of hot 1% agar in the desired Ringer's solution into culture dishes and wait until it is completely solidified. Fill the dishes about three-quarters full with the desired Ringer's solution. Agar-coated dishes help to prevent the embryos from sticking to the dish.
16. Micromanipulator: A simple manual micromanipulator works well for cell transplantation (e.g., Narishige, MM-3).
17. Watchmaker's forceps.
18. Sharpened glass needle: The end of a Pasteur pipet is pulled to a fine tip on a small gas burner or spirit lamp.
19. Blunt glass needle: Burn the tip of sharp glass needle for a while.
20. Tungsten needle: sharpened from a fine tungsten filament (0.2 mm in diameter, e.g., Nilaco Corp., Tokyo). To sharpen, mount into a Pasteur pipet or needle holder, then insert repeatedly in the side of a very hot flame; further sharpen by repeatedly soaking the tip of the filament in melted sodium nitrite. For melting, heat the crystal in a quartz melting pot with a gas burner. Do not use ceramic pots, which cannot withstand the heat of melting sodium nitrite. This process is very dangerous and great care should be taken.
21. Mold for making holes in agar-coated dishes (**Fig. 1A**): agar-coated dishes containing multiple holes are required for holding embryo/yolk cell combinations to ensure complete adhesion between the donor and host tissues. The holes in the agar should just fit the recombinants. The best diameter for the hole is approximately 1.2 mm. To make these dishes, we use a silicone rubber mold. The silicone mold is made by pouring liquid silicone mixed with a hardener onto a stainless plate containing holes ($\varnothing$ = 1.2 mm), in which one end of the hole has been sealed with tape (**Fig. 1B**).

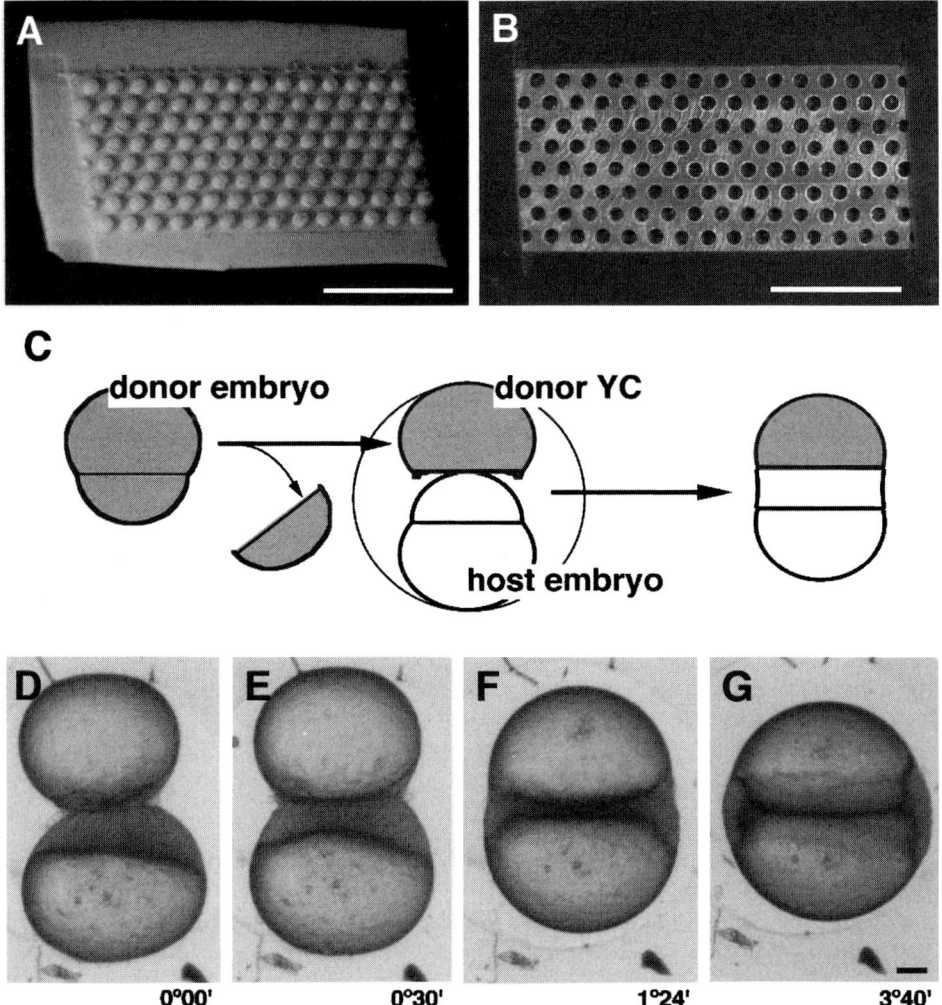

Fig. 1. Transplantation of the yolk cell. (**A**) A silicone rubber mold for agar holes. Scale bar = 10 mm. (**B**) A stainless steel plate used for production of the silicone mold shown in A. The diameter of the hole is 1.2 mm. (**C**) Schematic representation of the experiment. (**D–G**) The process of adhesion between the donor yolk cell (upper) and the host embryo (lower) which are kept in an agar hole. Scale bar = 100 μm.

22. Hooked glass needle (**Fig. 2A**) used for removal of the yolk mass: Glass capillaries are pulled to fine tips on an electrode puller. The tips are then fire polished with a microforge. To make a hooked shape, heat the center of the pulled capillary with a microforge.

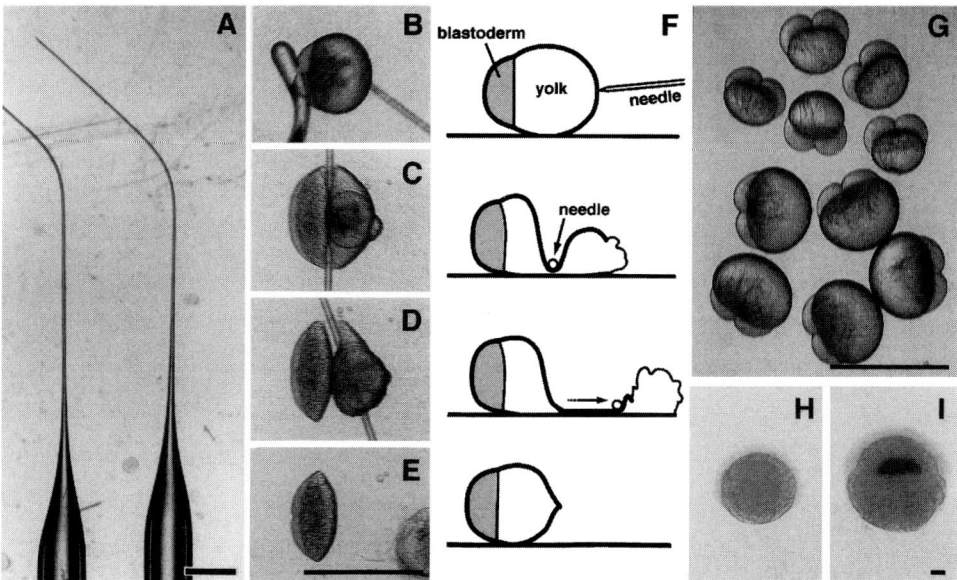

Fig. 2. Removal of the vegetal yolk hemisphere. (**A**) Hooked glass needles used in the operation. (**B–E**) The process of the operation. The vegetal yolk mass is squeezed out though a small hole made in the vegetal yolk membrane. The operation should be finished in a few seconds. (**F**) Schematic representations of the operation shown in **B–E**. (**G**) Two-cell stage embryos. As compared with normal embryos (lower five), the experimantally manipulated embryos (upper five) are smaller in size but undergo a normal cleavage. (**H,I**) *In situ* hybridization with *goosecoid* probe at the 50% epiboly stage. The manipulated embryo does not express *goosecoid* (**H**) whereas the control embryo (**I**) shows a positive signal in the future dorsal region. Scale bar = 1 mm (A–G), 100 μm (H,I).

3. Methods
3.1. Transplantation of Yolk Cell: Analysis of Mesoderm Induction

A schematic representation of the experiment described below is shown in **Fig. 1C**.

1. Label donor embryos at the 1–8 cell stages: inject a rhodamine–biotin mixture (1.65% rhodamine–dextran and 1% biotin–dextran in 0.2 M KCl) into the yolk (microinjection into zebrafish embryos, *see* **Chapter 11**). The injected dye spreads through intercellular cytoplasmic connections to all cells of the blastoderm. This ensures that the cells used for transplantation are labeled, and hence recognizable from those of the host embryos.

2. Preparation of agar holes: pour the appropriate amount of hot 1.5% agar in 1X Ringer's solution into culture dishes and immediately place the silicon mold (*see* **item 21** under **Subheading 2.**) onto the hot agar. When the agar is completely solidified, carefully remove the mold and fill the agar-holed dish with 1X Ringer's solution (referred to as an "agar-hole dish").
3. Dechorionate labeled donor and host embryos (removing embryos from their chorions, *see* **Chapter 11**). Wash them three times with fresh (1/3)X Ringer's solution, transfer dechorionated donor or host embryos with a Pasteur pipet into agar-coated dishes containing (1/3)X Ringer and agar-hole dishes containing 1X Ringer, respectively.
4. Preparation of donor yolk cells: Donor yolk cells are usually prepared from midblastula embryos (1000 cell stage to sphere stage). Place labeled donor embryos in an agar-coated dish containing calcium-free (1/3)X Ringer's solution. Remove the blastoderm cells from the yolk cell mechanically with a sharpened glass needle. Gently pipet isolated yolk cells up and down in order to remove marginal cells that are tightly attached to the yolk cell. Make sure that most of the blastoderm cells have been removed (*see* **Note 1**). Carefully transfer isolated yolk cells to the agar-hole dish containing host embryos in 1X Ringer.
5. Before transplanting the yolk cell, make a small incision in the enveloping layer of the animal-pole region of the host embryo with a sharpened glass needle. This helps rapid adhesion between the donor and host tissues. Transplantation should then be carried out immediately. By use of a blunt glass needle, push both donor the yolk cell and the host embryo into a hole made in the agar, with the donor's yolk syncytial layer facing the host animal pole. Let the recombined embryos sit for about 30 min in 1X Ringer's solution, during which time the host blastoderm cells start to cover the donor yolk cell (**Fig. 1D–G**). The higher salt concentration in an agar-hole dish helps the manipulated embryos to heal, but it needs to be exchanged to a lower-salt-concentration (1/3)X Ringer's solution before the onset of epiboly.
6. Thirty minutes after the operation, replace 1X Ringer's solution with (1/3)X Ringer's solution by washing three times with (1/3)X Ringer's, taking care that the recombinants do not come out of their holes. Incubate them until they reach the appropriate developmental stage.
7. The recombinants may then be then fixed and examined for gene expression. For example, ectopic expression of mesodermal genes such as *no tail* and *goosecoid* is induced in the host cells around the grafted yolk cell *(2)*. It is difficult to culture these recombined embryos beyond the bud stage, probably due to a shortage of the cell number required for formation of two body axes (*see* **Notes 2–5**).

3.2. Removal of the Vegetal Yolk Mass: Production of Ventralized Embryos

A schematic representation of the method described next is shown in **Fig. 2B–F**.

1. Preparation of egg albumen: stir egg albumen with an eggbeater to make it dissolved easily. Leave it overnight at 4°C and use this liquefied egg albumen as a 100% concentration.
2. Prepare embryos by in vitro fertilization as described in *(7)*.
3. Transfer the fertilized embryos to an agar-coated dish containing 1X Ringer (without albumen). To produce ventralized embryos at a high frequency, the operation should be carried out within 30 min. of fertilization (*see* **Note 6**).
4. Soon after fertilization (5–10 min), yolk-free cytoplasm begins to segregate to the animal pole. Locate the vegetal end of the embryos. Stick the tip of a hooked glass needle into the vegetal yolk membrane (**Fig. 2B**).
5. Place the hooked glass needle in the equatorial region of the yolk mass. Gently push the needle, trying to squeeze the vegetal yolk mass out of the embryo (**Fig. 2C**). For complete removal, move the needle slowly toward the vegetal end while applying continuous pressure against the agar bed (**Fig. 2D**).
6. Let the operated embryo sit for a few minutes. The operated embryos resume a round shape and start to recover from the damage to the yolk membrane. (**Fig. 2E,F**).
7. Transfer these manipulated embryos to an agar-coated dish containing 1X Ringer's supplemented with 1.6% egg albumen.
8. If culture of the embryos for an extended period is required replace the 1X Ringer's with (1/3)X Ringer's without albumen at 50% epiboly.
9. Fix the embryos at the appropriate developmental stage and examine gene expression. For example, these manipulated embryos show no *goosecoid* mRNA expression at the onset of gastrulation (**Fig. 2H,I**) whereas *no tail* is normally expressed in the germ ring (*see* **Note 7**).

3.3. Transplantation of Organizer Tissues: Analysis of Neural Induction

3.3.1. Transplantation of the Embryonic Shield

A schematic representation of the experiment described below is shown in **Fig. 3**.

1. Label donor embryos at the 1–8 cell stages by injecting a rhodamine–biotin mixture (1.65% rhodamine–dextran and 1% biotin–dextran in 0.2 M KCl) into the yolk.
2. Dechorionate the labeled donor and host embryos. After washing three times with fresh (1/3)X Ringer's, transfer dechorionated embryos with a Pasteur pipet into agar-coated cultured dishes containing (1/3)X Ringer's. Incubate them (at 28.5°C) until use.
3. Place a shield-stage donor embryo into the well of a depression slide containing PBS. Then, 2% methyl cellulose in (1/3)X Ringer's is spread on the surface of the well to hold the embryo, which is then overlaid with a drop of PBS (**Fig. 3A**).

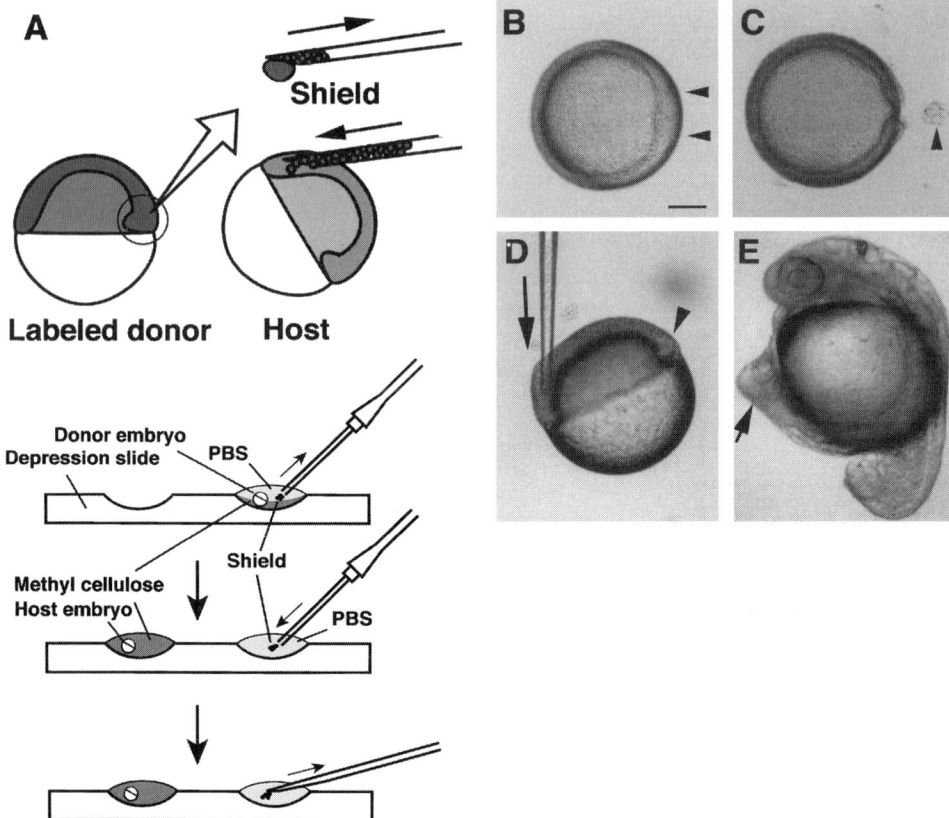

Fig. 3. Transplantation of the embryonic shield. (**A**) Schematic representation of the experiment. (**B**) Animal-pole view of a shield-stage embryo (6 h). The shield region (thickened germ ring) is indicated by a pair of arrowheads. (**C**) Animal-pole view of the shield-stage embryo in which the embryonic shield has been removed, the arrowhead indicates the isolated shield tissue. (**D**) The host embryo (shield-stage) into which is inserted on the ventral side the micropipet containing donor tissue. The arrowhead indicates the host shield region. (**E**) The secondary axis with anterior head structures (arrow) induced by the transplanted shield in a 20-h host embryo. Scale bar = 100 μm.

4. Prepare another depression slide for transplantation (transplantation slide). It is better to use a depression slide containing two wells (**Fig. 3A**). Fill one of the wells with 2% methyl cellulose in (1/3)X Ringer's for the host embryo and the other with PBS for the donor tissues. Place a host embryo (dome to shield stage) into the well containing 2% methyl cellulose in (1/3)X Ringer's.
5. Under a dissecting microscope, isolate the embryonic shield by cutting the embryo with a sharpened tungsten needle while the embryo is being held by a

watchmaker's forceps (**Fig. 3B,C**). Make sure that isolated tissues are free of yolk if the yolk membrane is damaged.

6. Transfer the shield tissue to the well of the transplantation slide containing the host embryo with a capillary glass (Narishige, G-1) equipped with a rubber aspirator tube to the mouth.
7. Place the transplantation slide on the stage of a microscope equipped with a micromanipulator. It is best if the microscope has a fixed stage; otherwise, the micromanipulator will need to be mounted on the stage. The operation can be carried out under a dissecting microscope if high magnification (X40–X60) is available.
8. Position a glass micropipet with a broken tip near the dissected shield under the objective and pipet up a little of the PBS solution. Try to keep zero pressure at the tip of the micropipet.
9. Suck the cells gently from the shield tissue into the micropipet.
10. Withdraw the micropipet and move the slide or stage so that the micropipet is now located next to the host embryo, while watching under the objective.
11. Insert the micropipet into the appropriate position of the host embryo, on the ventral side if the shield is visible (**Fig. 3A,D**). Do not damage the yolk cell membrane (*see* **Notes 8–12**).
12. Expel the cells with gentle pressure.
13. Withdraw the micropipet from the host embryo.
14. Add gently a small aliquot of (1/3)X Ringer's to the well containing the host embryo.
15. Place the slide containing the hosts in a plastic culture dish ($\varnothing$ = 10 cm) and incubate it. You may pour 10 to 20 mL of (1/3)X Ringer's gently into the dish so as to completely cover the slide.
16. After a few hours' incubation, the methyl cellulose solution becomes less viscous and the host embryos become detached from the bottom of the depression slides. Transfer them carefully with a Pasteur pipet to a culture dish containing fresh (1/3)X Ringer's and incubate them for an appropriate period. The secondary axis becomes visible during the late gastrula to 24-h stages (**Fig. 3E**).

3.3.2. Transplantation of COS7 Cells Secreting Organizer Factors

A schematic representation for the experiment described below is shown in **Fig. 4**. For making cell aggregates of COS7 cells, we essentially follow the protocol described elsewhere *(9)*.

1. Three days before the transplant will take place, plate COS7 cells (approximately 5×10^5) on a small culture dish ($\varnothing$ = 35 mm) so that they will be 70–80% confluent on the next day. The culture medium used is Dulbecco's modified Eagle medium (DMEM) supplemented with 10% fetal calf serum (FCS).
2. Two days before the operation. Transfect the cells with plasmid DNAs with Lipofectamine™ following the manufacturer's protocol. Briefly, 12 µL of Lipofectamine™ and 2 µg of plasmid DNA (purified by cesium chloride banding) are diluted separately into 100 µL of aliquots of serum-free DMEM (with-

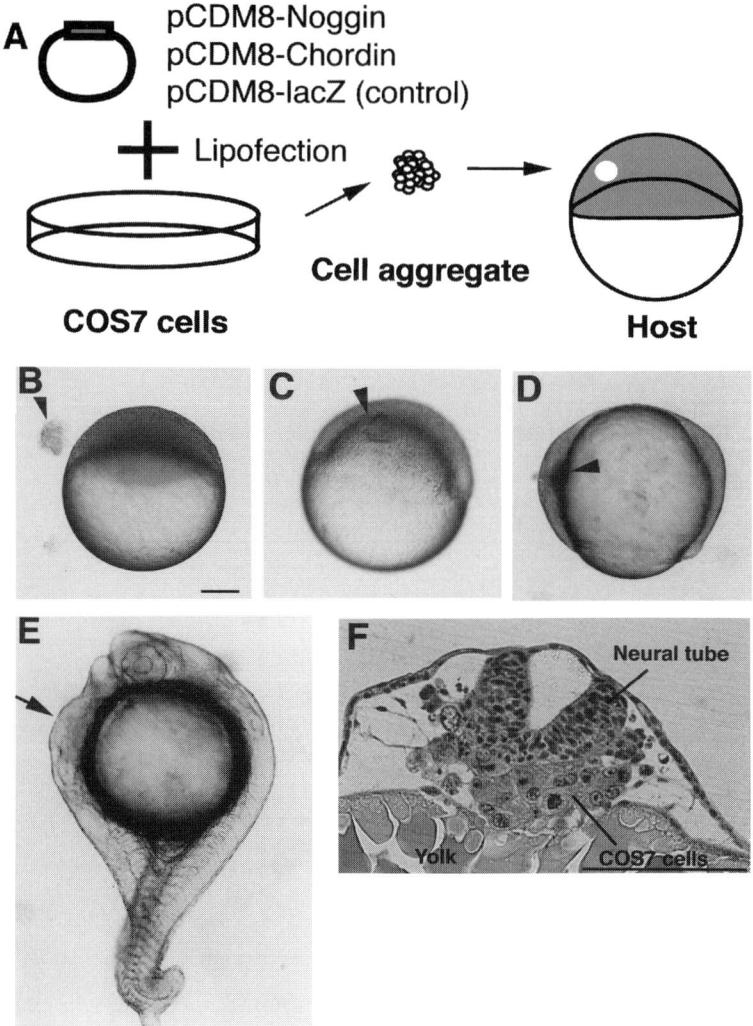

Fig. 4. Transplantation of COS7 cells secreting organizer factors. (**A**) Schematic representation of the experiment. (**B**) The cell aggregate (arrowhead) placed near the host embryos (dome stage, $4^{1}/_{3}$ h). (**C**) The host embryo (50%-epiboly, $5^{1}/_{4}$ h) grafted with the cell aggregate (arrowhead, about 1 h after transplantation). (**D**) The host embryo (80% epiboly, 8 h) grafted with the cell aggregate (arrowhead, about 8 h after transplantation). The ventral epiblast around the graft becomes thick, indicating neural plate formation on the ventral side. (**E**) Secondary axis (arrowhead) induced by Noggin/Chordin COS7 at 24 h. The secondary axes induced by COS7 tend to show a cyclopic phenotype (one-eyed head), probably because of a lack of axial mesoderm. (**F**) Cross sections of the secondary axis at the level of the hindbrain. The COS7 cell mass is located under the induced neural tube. Scale bars = 100 μm.

out antibiotics). These aliquots are gently mixed and incubated at room temperature for 15 min to form complexes. The complexes are diluted with 0.8 mL of serum-free DMEM (without antibiotics) and the mixture (transfection medium) is added to subcofluent cells in a small culture dish ($\varnothing = 35$ mm). The cells are rinsed twice with serum-free DMEM (without antibiotics) prior to the addition of the diluted complexes. We use 1 µg each of pCDM8 containing *Xenopus* noggin and chordin cDNAs or 1–2 µg of pCDM8 containing *lacZ* cDNA as a control.
3. After a 6-h incubation in 1 mL of transfection medium, add 0.8 mL of DMEM and 0.2 mL of FCS to the dish and incubate overnight.
4. On the morning of the day before the transplant, change the medium to fresh DMEM/10% FCS. In the evening, harvest the cells and replate them on a culture dish coated with 1% agar. Incubate them overnight in DMEM/10% FCS. To make the agar dish, pour 0.5–0.6 mL of hot 1% agar in distilled water or PBS into a small culture dish ($\phi = 35$ mm) and wait until completely solidified.
5. On the morning of the day of the transplant, make sure that cell aggregates are formed. The size of the cell aggregates depends on the density of the cells plated and/or to what degree they are dissociated at the stage of plating.
6. Dechorionate host embryos (sphere to shield stage) and place them in the well of a depression slide containing 2% methyl cellulose in (1/3) X Ringer's.
7. Transfer a small group of cell aggregates from the culture dish into the well containing the host embryos, using a glass capillary equipped with an aspirator tube or a Pasteur pipet.
8. Under a dissecting microscope, pick up a cell aggregate of the appropriate size or cut out a small piece from a bigger aggregate with a sharpened tungsten needle (approximately 50 µm in diameter is preferable). Move the aggregate near the host using the needle (**Fig. 4B**).
9. Make a small incision (*see* **Note 11**) in the enveloping layer of the host with a sharpened needle. Insert the cell aggregate into the deep cell layer using the needle, taking care not to damage the yolk membrane (**Fig. 4C**). It is essential to make the incision as small as possible, otherwise the cell aggregate will be pushed out during epiboly.
10. Add gently a small aliquot of (1/3)X Ringer's to the well containing the host embryos.
11. Place the slide in a plastic culture dish ($\varnothing = 100$ mm) and incubate it. You may pour 10–20 mL of (1/3)X Ringer's gently into the dish so as to completely cover the slide.
12. After a few hours' incubation, the methyl cellulose solution becomes less viscous and the host embryos detach from the bottom of the depression slides. Transfer them carefully with a Pasteur pipet into fresh (1/3)X Ringer's and incubate them for an appropriate period. The secondary axis becomes visible during late gastrula to 24-h stages (**Fig. 4D–F**) (*see* **Note 12**).

4. Notes

1. During all procedures of operation, make sure that dechorionated embryos and isolated tissues do not touch the surface of any solution or they will burst because of the surface tension of the liquid.

2. If the donor tissues are labeled with biotin–dextran, they are visualized in the host by biotin–peroxidase staining as described elsewhere *(7,10)*.
3. Yolk cells prepared from 512-cell to sphere stage embryos show the same inducing activity.
4. Under our experimental conditions, we cannot remove the marginal cells completely with the yolk cell intact, probably due to the tight adhesion of the marginal cells with the yolk syncytial layer. Thus, the yolk cells to be transplanted contain a few marginal cells. It is confirmed that a few marginal cells attached to donor yolk cells do not affect gene expression in the host cells *(2,14)*.
5. Although the higher salt condition (1X Ringer's) is required for wound healing, replacement of 1X Ringer's with low-salt (1/3)X Ringer's is essential for successful transplantation. However, the timing of replacement differs for each experiment and even on batches of eggs. It is best to carry out the replacement either as soon as firm adhesion between the donor and host tissues is established, or when the manipulated embryos recover from their damage. It is known that the higher salt condition perturbs the gastrulation process in dechorionated zebrafish embryos.
6. Originally, this technique was developed for goldfish embryos **(5,11)**. In these experiments, the embryos were bisected from the vegetal yolk hemisphere using nylon fibres crossing the equator. This method is only applicable for zebrafish embryos until approximately 15 min postfertilization, because the yolk membrane loses its softness after this stage. However, the modified method described here can be applied to zebrafish embryos at any developmental stage.
7. The embryos from which the vegetal yolk mass has been removed make no dorsal structures, such as notochord, somites, and neural tube. The frequency of abnormality decreases as the age at which the vegetal yolk hemisphere is removed increases *(5)*. For zebrafish embryos, the frequency of a ventralized phenotype is highest when yolk mass removal is carried out 20 min after fertilization and no ventralized embryos are obtained by this manipulation after the 8-cell stage *(14)*.
8. To avoid damage to the yolk membrane during shield transplantation, it is better to perform the injection by moving the stage. Once the position of the micropipet is fixed under the objective, we never touch the micromanipulator during transplantation.
9. Methyl cellulose, when contaminating the deep cell layer, inhibits normal development of the embryo, especially the process of epiboly. Thus, it is essential not to take up the methyl cellulose solution into the micropipet before insertion into the embryo. Similarly, too much PBS injected into the embryo leads to abnormal development. Try to transplant tissues or cells with as little medium as possible.
10. During the process of transplantation, the embryonic shield tends to disintegrate into small fragments or even single cells, because of weak cell adhesion at this stage. Because the inducing ability of the shield is displayed to the full when transplanted as a tissue mass, it is important to handle the shield tissue gently, taking care to avoid dissociation.

11. To obtain a secondary axis with anterior head structures, it is essential to graft organizer tissues halfway between the blastoderm margin and the animal pole. When the inducing tissues are grafted near the blastoderm margin, secondary axes are frequently induced, but those axes lack anterior head structures *(12,13)*.
12. The fish organizer (embryonic shield) and mammalian COS7 cells transfected with *Xenopus* noggin and chordin cDNAs (Noggin/Chordin COS7) induce secondary axes equally when transplanted at mid-blastula to early gastrula stage on the ventral side of the fish embryo (**Fig. 3E** and **Fig. 4E**). However, these inducing tissues behave differently in terms of their contribution to the secondary axes produced. Grafted embryonic shield contributes to the axial mesoderm and the ventral part of the neural tube *(13)*, whereas the Noggin/Chordin COS7 shows no sign of self-differentiation but is present in a cell mass under the neural tube (**Fig. 4F**; *15*). No axial mesoderm is detectable in the secondary axis induced by the Noggin/Chordin COS7.

Acknowledgments

We would like to thank Dr. Etsuro Yamaha for critical advice during the development of the yolk cell bisection and transplantation method and Professor Atsushi Kuroiwa for supporting our study. This work was supported in part by grants-in-aid from the Ministry of Education, Science, and Culture of Japan, by CREST (Core Research for Evolutional Science and Technology) of Japan Science and Technology Corporation (JST), and by research funds of the Asahi Glass Foundation and Naito Foundation.

References

1. Long, W. L. (1983) The role of the yolk syncytial layer in determination of the plane of bilateral symmetry in the rainbow trout, *Salmo gairdneri* Richardson. *J. Exp. Zool.* **228,** 91–97.
2. Mizuno, T., Yamaha, E., Wakahara, M., Kuroiwa, A., and Takeda H. (1996) Mesoderm induction in zebrafish. *Nature* **383,** 131–132.
3. Scharf, R. R. and Gerhart, J. C. (1983) Axis determination in eggs of *Xenopus laevis*: a critical period before first cleavage, identified by the common effects of cold, pressure and ultraviolet irradiation. *Dev. Biol.* **99,** 75–87.
4. Strähle, U. and Jesuthasan, S. (1993) Ultraviolet irradiation impairs epiboly in zebrafish embryos: evidence for a microtubule-dependent mechanism of epiboly. *Development* **119,** 909–919.
5. Mizuno, T., Yamaha, E., and Yamazaki, F. (1997) Localized axis determinant in the early cleavage embryo of the goldfish, *Carassius auratus. Dev. Genes Evol.* **206,** 389–396.
6. Gerhart, J., Danilchik, M., Doniach, T., Roberts, S., Rowning, B., and Stewart, R. (1989) Cortical rotation of the *Xenopus* egg: consequences for the anteroposterior pattern of embryonic dorsal development. *Development* **107(Suppl.),** 37–51.
7. Westerfield, M. (1993) *The Zebrafish Book: A Guide for the Laboratory Use of Zebrafish* (*Danio rerio*), University of Oregon Press, Eugene, OR.

8. Yamaha, E. and Yamazaki, F. (1993) Electrically fused-egg induction and its development in the goldfish, *Carassius auratus*. *Int. J. Dev. Biol.* **37,** 291–298.
9. Tonegawa, A., Funayama, N., Ueno, N., and Takahashi, Y. (1997) Mesodermal subdivision along the mediolateral axis in the chicken controlled by different concentrations of BMP-4. *Development* **124,** 1975–1984.
10. Miyagawa, T., Amanuma, H., Kuroiwa, A., and Takeda, H. (1997) Specification of posterior midbrain region in zebrafish neuroepithelium. *Genes Cells* **1,** 369–377.
11. Tung, T. C., Chang, C. Y., and Tung, Y. F. Y. (1945) Experiments on the developmental potencies of blastoderms and fragments of Teleostean eggs separated latitudianally. *Proc. Zool. Sci.* **115,** 175–188.
12. Hatta, K. and Takahashi, Y. (1996) Secondary axis induction by heterospecific organizers in zebrafish. *Develop. Dynam.* **205,** 183–195.
13. Shih, J. and Fraser, S. E. (1996) Characterizing the zebrafish organizer: microsurgical analysis at the early shield stage. *Development* **122,** 1313–1322.
14. Mizuno, T., Yamaha, E., Kuroiwa, A., and Takeda, H. (1999) Removal of vegetal yolk causes doral deficiencies and impairs doral-inducing ability of the yolk cell in zebrafish. *Mech. Dev.*, in press.
15. Koshida, S., Shinya, M., Mizuno, T., Kuroiwa, A., and Takeda, H. (1998) Initial anteroposterior pattern of the zebrafish central nervous system is determined by differential competence of the epiblast. *Development* **125,** 1957–1966.

3

Ribonuclease Protection Analysis of Gene Expression in *Xenopus*

Craig S. Newman and Paul A. Krieg

1. Introduction

When characterizing the developmental expression of a novel gene, or when examining the response of a known gene to experimental manipulations, it is important to be able to assay mRNA transcript levels accurately. Although a number of techniques for transcript analysis are available, one of the most useful and widespread is the ribonuclease (RNase) protection assay (for example, *see* **Fig. 1**). The major advantages of RNase protection analysis are good sensitivity, excellent specificity, and the linear response to differing transcript levels. Perhaps the major disadvantage of RNase protection is the need to prepare special RNA probes and the fact that the RNA samples used for RNase protection analysis are destroyed and therefore cannot be reused. In addition to expression analysis, RNase protection can also be applied to a number of additional experimental goals, including mapping of transcription start sites, mapping of intron/exon boundaries, analysis of alternative splicing, and determination of the rate of degradation of nucleic acids introduced into the embryo.

The basic premise of the RNase protection assay is as follows. A short, radioactively labeled RNA probe, complementary to the desired target sequence, is produced in an in vitro transcription reaction and added to an heterogeneous sample of RNAs. The probe then hybridizes to target transcripts, forming double-stranded RNA duplexes. These double-stranded RNA regions are resistant to degradation by most ribonucleases. Therefore, a mixture of RNases is used to digest both the unhybridized sample RNA and the excess radiolabeled probe, leaving RNA duplexes intact. After inactivation of the RNase by a combination of protease digestion and phenol-chloroform extraction, the protected RNA probe is fractionated on a polyacrylamide gel and detected by autoradiography.

From: *Methods in Molecular Biology, Vol. 127: Molecular Methods in Developmental Biology: Xenopus and Zebrafish* Edited by: M. Guille © Humana Press Inc., Totowa, NJ

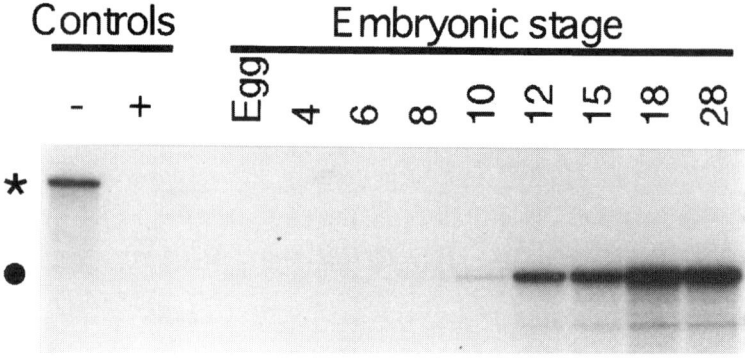

Fig. 1. Developmental profile of the EF-1α gene transcript as seen by a 70-min exposure of the final acrylamide gel. The input probe is denoted by an asterisk; the darkened circle marks the level of the protected fragment. Expression can first be detected at stage 10. Note that EF1-α is an exceptionally abundant transcript and most gene products will require a significantly longer exposure time to visualize the protected fragments.

It is perhaps worth comparing RNase protection analysis with the other commonly used transcript analysis techniques in a little more detail. First, the RNase protection assay is significantly more sensitive (estimated at 8- to 10-fold) than methods, like Northern blotting or dot blotting, that fix RNA to a solid support *(1,2)*. This is most likely because both the RNA probe and the target sequence are free in solution and, therefore, available for hybridization. In contrast, binding to a membrane is believed to make a large proportion of the target RNA inaccessible for hybridization. Second, the RNase protection assay is exceptionally useful for distinguishing between closely related genes. As the protection probes are usually rather short (typically several hundred nucleotides), they can be directed to an area of the gene most dissimilar to other family members. In practice, very small differences, sometimes corresponding to a single base mismatch can be detected using RNase protection *(3)*. Third, because only a small portion of each target RNA will actually hybridize to the labeled probe, the RNase protection assay can tolerate some degradation of the RNA sample before the results are compromised. Lee and Costlow *(2)* have found that RNA sheared to as small as 400–500 base pairs is still suitable for use in the RNase protection assay. This is not the case with Northern blot analysis, where any degradation of the input RNA results in a loss of sensitivity and clarity. The major disadvantages relative to Northern blotting are that RNase protection destroys the target RNA during the digestion reaction and that the assay does not provide any information about the size of the original transcript or, in general, about the presence of alternatively spliced transcripts. This information is most appropriately obtained using RNA blot procedures *(4)*.

Another commonly used method for expression analysis is reverse transcription of mRNA coupled to the polymerase chain reaction (RT-PCR). This technique has the advantage of extreme sensitivity; however, the nonlinearity of PCR, particularly after a large number of amplification cycles, makes this method at best only semiquantitative and the results generated can often be misleading. In contrast, because the RNase protection assay utilizes a molar excess of probe relative to target RNA, the majority of the target molecules are detected, resulting in a quantitative method for estimating RNA abundance *(2)*.

2. Materials
2.1. Isolation of RNA from Frog Embryos

1. Buffer A: 50 mM Tris-HCl, pH 7.5, 50 mM NaCl, 10 mM EDTA, 0.5% sodium dodecyl sulfate (SDS).
2. Proteinase K: 25 mg/mL stock in stabilization buffer (10 mM Tris-HCl, pH 8.0, 50 mM KCl, 1.5 mM MgCl$_2$, 0.45% Nonidet P-40, 0.45% Tween-20).
3. 8 M LiCl.

2.2. RNA Probe Synthesis and Purification

1. Template DNA: linearized, phenol–chloroform extracted, ethanol precipitated and resuspended at a final concentration of about 1 mg/mL in water.
2. Dithiothreitol (DTT): 100 mM.
3. Bovine serum albumin (BSA): 1 mg/mL.
4. Unlabeled ribonucleotide triphosphate mixture: the three unlabeled nucleotides at a final concentration of 5 mM each.
5. Ribonuclease inhibitor.
6. Radiolabeled ribonucleotide triphosphate: [α-^{32}P]-cytidine 5'-triphosphate (CTP) or uridine 5'-triphosphate (UTP) at about 800 Ci/mmol and 10 µCi/µL. For convenience, we will assume the use of CTP throughout these protocols; however, *see* **Note 1** and **ref. *1*** for more information.
7. Unlabeled ribonucleotide triphosphate: 200 µM solution of CTP. This reagent is generally not necessary, except for the preparation of low specific activity control probes for the detection of abundant sequences, or for the preparation of very long probes.
8. 10X transcription buffer: 400 mM Tris-HCl, pH 7.5, 60 mM MgCl$_2$, 20 mM Spermidine (Sigma, St. Louis, MO). We have found that this transcription buffer has a limited life-span and that, in general, transcription buffer older than about 1 mo should not be used.
9. Bacteriophage RNA polymerases (10 U/µL or higher). Use T7, T3, or SP6 RNA polymerase, depending on the vector and the orientation of the insert.
10. DNase I: RNase-free.

11. Formamide gel loading buffer: formamide containing 0.1% w/v xylene cyanol, 0.1% Bromophenol Blue, 10 mM EDTA.
12. Probe elution buffer: 500 mM NH$_4$OAc, 10 mM MgCl$_2$, 1 mM EDTA, 0.1% SDS.

2.3. Hybridization and RNase Digestion

1. Target RNA: stored as an ethanol precipitate.
2. Formamide: molecular biology grade.
3. 10X hybridization buffer: 4 M NaCl, 400 mM PIPES, pH 6.4, 10 mM EDTA.
4. RNase digestion buffer: 300 mM NaOAc, 10 mM Tris-HCl, pH 7.5, 5 mM EDTA.
5. RNase A/T1 mixture: RNase A at 500 U/mL and RNase T1 at 20,000 units/mL (e.g., Ambion RNase cocktail, cat. no. 2286, Ambion, Austin, TX). Note that 500 U/mL of RNase A is approximately equivalent to 0.75 mg/mL.
6. Proteinase K: 25 mg/mL in stabilization buffer (see above for composition).
7. SDS: 10% (w/v) solution.
8. Carrier RNA: 10 mg/mL solution of yeast RNA resuspend in Tris-EDTA (TE) (10 mM Tris–HCl, pH 8.0, 1 mM EDTA).

3. Methods
3.1. Isolation of RNA from Embryos (see Note 2)

1. Homogenize embryos in buffer A by rapidly pipeting up and down with a pipetman and by vigorous vortexing. A maximum of 20 embryos should be processed per milliliter of buffer A, otherwise some degradation of RNA may be observed (*see* **Note 3**).
2. Following homogenization, add a 1/100th volume of proteinase K stock solution (final concentration 0.25 mg/mL) and incubate for 1 h at 37°C (*see* **Note 4**).
3. Extract the homogenate twice with phenol–chloroform and precipitate the aqueous phase by addition of 2.5 volumes of ethanol and 1/10th volume of NH$_4$OAc (*see* **Note 5**). The pellet at this stage is rather large and has a waxy appearance, as a result of the presence of contaminating glycoproteins that are not removed by phenol extraction. In extractions from later-stage embryos, the pellet will also contain quite large amounts of genomic DNA. Both the glycoproteins and the DNA can be removed by a LiCl precipitation step.
4. Following centrifugation, resuspend the RNA pellet in 400 µL of TE and mix with an equal volume of 8 M LiCl.
5. After incubation for 2 h on ice (or overnight at 4°C), recover the RNA by centrifugation for 10 min in a microcentrifuge.
6. Resuspend the pellet in 10 µL of TE per embryo. Note that the RNA pellet after LiCl precipitation is relatively difficult to resuspend compared to an ethanol precipitation pellet. Store the RNA at –20°C as an ethanol precipitate, after addition of 1/10th volume of NH$_4$OAc and 3 volumes of ethanol. In this case, one embryo equivalent of total RNA is stored in 40 µL total volume. Alternatively, the RNA solution in TE may be stored at –80°C (*see* **Note 6**).

3.2. RNA Probe Synthesis and Purification (see Note 7)

1. Assemble the components of the probe synthesis reaction in the order given below. To avoid possible precipitation of the template DNA by the spermidine in the transcription buffer, the reaction should be assembled at room temperature.

Linear template DNA + H$_2$O (1 µg total)	2 µL
DTT	1 µL
BSA	1 µL
Unlabeled nucleotide mix (adenosine 5'-triphosphate [ATP], guanosine 5'-triphosphate [GTP], UTP)	1 µL
Ribonuclease inhibitor	1 µL
[α-^{32}P] CTP (see **Notes 1** and **8**)	2.5 µL
10X transcription buffer	1 µL
Bacteriophage RNA polymerase (5-10 units)	0.5 µL
Total volume	10 µL

2. Incubate the assembled reaction mix at 37°C, *or preferably at a lower temperature*, for 1 h (*see* **Note 9**).
3. Following the transcription reaction, remove the template DNA by the addition of 1 µL of RNase-free DNase I and incubation at 37°C for 15 min.
4. In general, it is probably necessary to gel purify the full-length RNA probe (*see* **Note 10**). To one-half of the probe synthesis reaction add an equal volume of formamide loading buffer and fractionate on a small 6% acrylamide denaturing gel (*see* **Note 11**).
5. When electrophoresis is complete, determine the position of the full-length probe by exposure to X-ray film for 30–60 s (*see* **Fig. 2**). Excise the portion of the gel containing the full-length transcript.
6. Add the slice of gel to 200 µL of elution buffer in an Eppendorf tube and incubate at 37°C.
7. After about 2 h of elution, remove a 1-µL sample in a pipet tip and estimate the amount of probe eluted using a hand-held monitor (*see* **Note 12**).
8. Remove the remainder of the eluted probe to a fresh tube and store as an ethanol precipitate by addition of 2.5 volumes of ethanol (salt is already present in the elution buffer).

3.3. Hybridization and RNase Digestion (see Note 13)

1. After thorough vortexing of the ethanol precipitated embryonic RNA samples (**Subheading 3.1., step 6**), aliquot an appropriate volume of each target RNA suspension into a fresh Eppendorf tube. In addition, two control tubes containing 20 µg of carrier RNA (again stored as an EtOH/NaOAc suspension) should be included in every set of reactions (*see* **Note 14**).
2. Using a hand-held monitor, determine the volume of probe suspension that contains 25–50 cps of labeled probe. Add this volume of RNA probe to each of the target RNAs and the two controls. A monitor reading of 50 cps corresponds to about 50,000 disintegrations/min (dpm) when measured using a scintillation counter. Also, add the probe for the control sequence if desired (*see* **Note 15**).

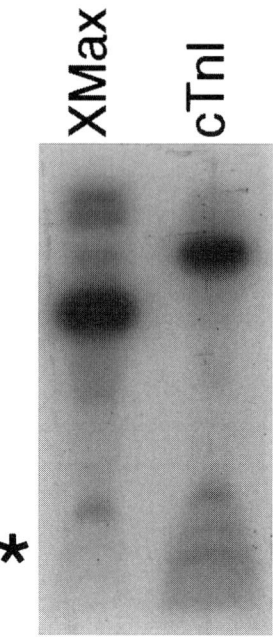

Fig. 2. Typical RNase protection probe synthesis. A 60-s exposure of newly synthesized probe produced by a 1-h incubation at 4°C. Note the high level of labeled nucleotide incorporation into full-length probe as indicated by the almost total lack of free nucleotide (denoted by the asterisk).

3. Centrifuge the samples in a microfuge for 15 min to precipitate both the target and probe RNA and remove the supernatant using a fine plastic pipet tip or a drawn-out glass pipet. Make sure that no liquid remains in the tube.
4. Resuspend the RNA pellet in 16 µL of formamide (*see* **Note 16**).
5. To each tube, add 2 µL of H$_2$O and 2 µL of 10X hybridization buffer and mix thoroughly.
6. Heat the hybridization solution to 100°C for 2 min to denature any secondary structure.
7. Incubate the hybridization reaction for at least 6 h at 45°C (*see* **Note 17**).
8. With the exception of one of the control samples, add 200 µL of RNase digestion solution to each tube. This is generally a 200:1 dilution of RNase A/T1 cocktail in RNase digestion buffer (*see* **Note 18**). To the remaining control tube, add 200 µL of digestion buffer without RNase.
9. Incubate at 37°C for 30 min.
10. RNase inactivation is accomplished by the addition of 10 µL of 10% SDS and 2 µL of Proteinase K solution, followed by incubation for 15 min at 37°C.
11. Extract all samples with an equal volume of phenol–chloroform.
12. Precipitate the aqueous layer, containing the protected RNA, by addition of 2.5

Ribonuclease Protection Analysis

volumes of ethanol and 10 µg of carrier RNA. Salt is already present in the digestion buffer.

13. Recover the RNA by centrifugation for 10 min in a microfuge and then resuspend the pellet in 5 µL of formamide loading buffer. Heat to 100°C for 2 min and then analyze the protected material by fractionation in a 6% acrylamide denaturing gel (see **Note 19**).
14. Detect the protected probe fragments by autoradiography.

4. Notes

1. Except in unusual circumstances, radiolabeled CTP or UTP should be used in the synthesis reaction (*1*). Where possible, a template containing homopolymeric stretches complementary to the labeled nucleotide should be avoided, as the limiting concentration of labeled nucleotide can result in a high proportion of incomplete transcripts. This problem is most commonly encountered when the extreme 3' end of a cDNA clone is chosen as the protection probe, and the presence of the poly-A tail represents a barrier to the use of labeled UTP. For most purposes therefore, we recommend the use of CTP as the radiolabeled nucleotide.
2. Although care should be exercised during all steps, we have not found it necessary to use many of the precautionary measures normally associated with RNA work. Although we routinely autoclave stock solutions, it does not seem to be necessary to diethylpyrocarbonate (DEPC) treat solutions or plastic and glassware. Furthermore, the presence of RNase inhibitor in the transcription reaction appears to provide ample protection against contaminating RNases.

    The method described is an exceptionally easy and efficient method for obtaining total RNA from *Xenopus* embryos, which when used in conjunction with a LiCl precipitation yields a very clean product. Using the proteinase K–SDS extraction procedure, we routinely isolate 3–5 µg of total RNA from each *Xenopus* embryo. As 1–2% of total RNA is poly(A)$^+$ messenger RNA, this yield corresponds to approximately 50 ng of mRNA per embryo.
3. Whereas young embryos are homogenized very easily, late tailbud and older embryos sometimes require more effort and will not always become completely disrupted. For these late stages, it may be necessary to use a Dounce homogenizer or a mechanical mixer (e.g., Polytron, Brinkman Instruments, NY) to achieve complete disruption. Alternatively, we have had good results using a guanidinium–isothiocyanate RNA isolation procedure such as those described for use with zebrafish embryos (*5*), for late tadpole stage embryos, and for preparation of RNA from adult tissues.
4. At this point, the RNA preparation may be stored at –20°C for many days, without detectable degradation.
5. Alternatively, the RNA can be precipitated using one volume of isopropanol. This smaller volume is convenient and appears to be quite suitable for most purposes. Note, however, that the use of isopropanol may lead to the selective loss of some very small RNA species from the total RNA preparation.
6. This protocol generally results in the isolation of high yields of intact total RNA.

However, if you are new to the technique, it may be prudent to assess the quality and yield of the RNA by gel electrophoresis. Half an embryo equivalent of RNA (about 2 μg) should be fractionated on a 1% agarose gel containing formaldehyde, according to standard techniques *(6)* and visualized by ethidium bromide staining. Ideally, both the large and the small ribosomal bands will be clearly visible and the larger band should appear slightly more intense than the lower band.

7. The following parameters should be considered when preparing DNA template for the in vitro transcription reaction.

   a. Most standard DNA preparations methods yield plasmid of sufficient quality for transcription. Remember when designing an RNase protection probe that a bacteriophage promoter sequence must be located 3' of the gene so that a probe complementary to the target RNA (i.e., an antisense probe) is synthesized.

   b. Ideally, the probe should be between 150 and 400 bases in length, although for abundant messages, a smaller probe may be used. In general, the longer the probe, the greater the signal due to an increased number of labeled nucleotides in the protected probe fragment. However, as the probe length increases, it becomes more difficult to produce full-length product.

   c. The template DNA should be linearized using a restriction enzyme which leaves either a blunt end or a 5' overhang, as bacteriophage RNA polymerases may initiate a low level of transcription from 3' overhangs *(7)*. If there is no alternative, the 3' overhang should be blunted using T4 DNA polymerase *(6)*. Remember, also, that the restriction enzyme does not have to cut at a unique site in the plasmid, so long as it does not separate the desired template region from the bacteriophage promoter. In no instance have we found it necessary to gel isolate the template DNA fragment.

   d. It is desirable for the probe to contain a significant stretch of vector sequence. This helps in distinguishing the specific protected probe from any undigested probe that may survive the assay procedure. During the hybridization reaction, the gene-specific portion of the probe forms a duplex with the target RNA, whereas the vector-specific sequences form a single-stranded tail. These single stranded tails are readily digested during the RNase treatment. Assuming a relatively large stretch of vector sequence (25 bases or greater), the protected fragment will have a noticeably different size than the input probe, allowing the experimenter to distinguish the protected band from undigested probe. This is a particularly important consideration, as we have found that some probes produce a doublet of protected bands—one being the predicted size, and the second is same size as the unprotected probe. In general, the larger band tends to be of a less intensity but increases in intensity proportionally to the amount of input RNA (*see* **Fig. 3**).

The method described for probe production is extremely reliable and routinely results in synthesis of full-length RNA probes. In the standard reaction, the final concentration of labeled nucleotide is about 3 μM (about 10 ng of CTP per 10-μL reaction). Typically, a large proportion of the label is incorporated into RNA, representing a total weight of RNA probe of about 20 ng. Kits for in vitro tran-

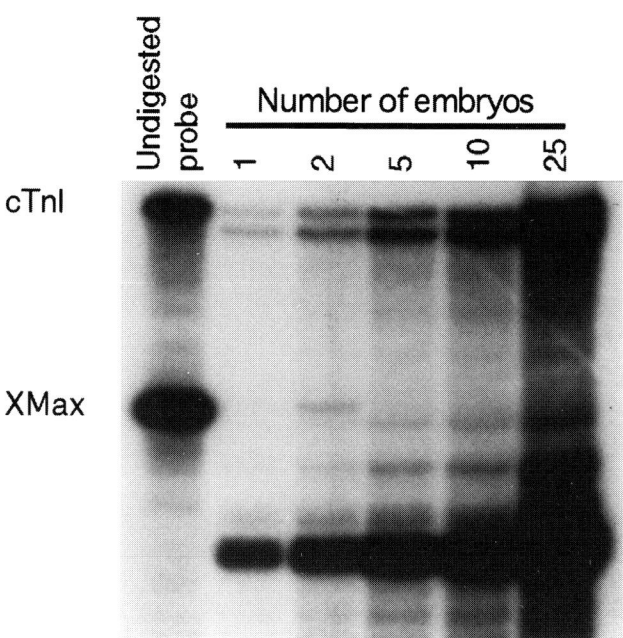

Fig. 3. Increasing amounts of target RNA results in increased intensity of protected bands. A 7-d exposure of an RNase protection using increasing amounts of input (tailbud stage) RNA. In the case of the cTnI probe, increasing amounts of target RNA results in an increased intensity of both the protected band as well as a larger band consistent with the size of the input probe.

scription are also available from a number of commercial sources and these represent a very efficient and convenient, if somewhat expensive, alternative to assembling your own reagents.
8. The majority of RNase protection experiments aim for maximum sensitivity of detection and therefore use a labeled probe at high specific activity. The standard reaction mix described above results in probe with a specific activity of approximately 10^9 dpm/μg. In some cases however, it may be necessary to supplement the labeled CTP with a small amount of unlabeled CTP (thereby increasing the concentration of the limiting nucleotide). For example, in the case of unusually long probes (greater than about 500 nucleotides) the proportion of full-length transcription products is increased by adding a small amount of unlabeled nucleotide. It may also be useful to reduce the specific activity of probes used to detect abundant control sequences, so that they do not overexpose the film during prolonged autoradiographic exposures. In either case, rather than using 2.5 μL of [α-^{32}P] CTP, use only 1.5 μL and then add 1 μL of 200 μM unlabeled CTP (bringing the final concentration of labeled nucleotide to about 20 μM). Under these conditions, the specific activity of the probe is closer to 10^8 dpm/μg.

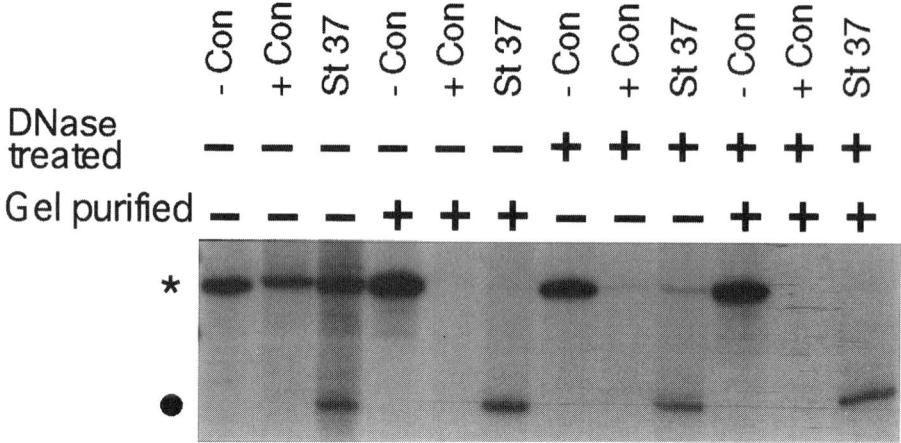

Fig. 4. Analysis of various probe purification options. An RNase protection for XMax using probe purified by various combinations of DNase I treatment and gel purification. Although all probes produce a protected band (denoted by the darkened circle), the lack of gel purification results in a shadow band at the size of the input probe (denoted by the asterisk). The minus and plus controls refer to the omission or presence, respectively, of RNase cocktail in the digestion mix.

9. Incubation at temperatures lower than 37°C results in a greater proportion of full-length transcripts, particularly when longer probes are being synthesized (8). A 4°C incubation is conveniently carried out in a standard laboratory refrigerator.
10. We have found that DNase treatment of the template alone, without subsequent gel purification, sometimes leaves enough contaminating DNA to produce a faint artifactual band on the final gel, running at the position of full-length probe (*see* **Fig. 4**). It is therefore important to gel isolate the full-length RNA probe. We purify our probes using a 6% acrylamide–urea gel with dimensions of 8 cm × 8 cm × 0.5 mm. These dimensions are generally considered to be more appropriate for protein gels, but fractionation is very quick and the resolution is quite adequate for isolation of full-length probe.
11. The remaining half of the reaction can be stored at –20°C for several days before use. During this time however, some autoradiolytic breakdown of the probe will occur (9); therefore, gel purification of the full-length probe is essential. Once gel-isolated, the probe should be used within 24 h and preferably immediately.
12. Approximately 50 cps of probe, as measured with a minimonitor, is required for each protection reaction. It is, therefore, a simple matter to multiply the counts in the 1-µL aliquot by the total volume of elution buffer in the tube and determine when sufficient probe has been eluted to carry out the experiment.
13. The method described for hybridization and digestion is based very closely on the originally described RNase protection protocol (*1,10*) and we have found it to be extremely reliable. A number of kits for RNase protection are now commer-

cially available. These kit protocols are faster and more convenient than the method outlined below, but, unfortunately, the recovery of the final protected probe can be somewhat unreliable.

Because RNA is soluble in formamide to very high concentrations, quite large amounts of total RNA can be used in this assay. We have found that greater than 75 μg of total RNA (equivalent to the RNA from about 25 embryos) will readily dissolve in 16 μL of formamide. As expected, the intensity of the protected band increases as the quantity of target RNA increases (see **Fig. 3**). In general, an overnight autoradiographic exposure will detect highly abundant transcripts in one embryo equivalent of RNA and most tissue-specific transcripts in about five embryos equivalent of RNA. Much rarer transcripts require more input RNA and a longer exposure time. For example, we have found it necessary to use total RNA from 15 embryos and a 2-wk autoradiographic exposure to adequately detect signal from a particularly rare homeobox gene sequence *(11)*.

14. One of these control tubes will be digested with RNase and the other will not. The first control ensures that intramolecular secondary structure within the probe does not result in a protected band, whereas the second control allows the integrity of the probe to be assessed after the entire protection protocol.
15. As there are steps in this protocol at which RNA may be lost, it is prudent to include an internal control. By including, in each reaction, a labeled probe for a ubiquitous transcript such as EF-1α or Max *(12,13)*, it is possible to evaluate the amount of RNA present in each sample (see **Fig. 2**). This also allows the estimation of relative transcript abundance between samples by comparison to the loading control. As most of these ubiquitous control transcripts are very abundant, it may be necessary to reduce the specific activity of the control probe by including unlabeled competitor nucleotide during the synthesis reaction (see **Note 8**). This has the effect of reducing the overall signal from the control to a more manageable level.
16. Resuspension should occur fairly rapidly, but heating may be used if necessary.
17. The hybridization reaction is carried out at 45°C. Both higher and lower temperatures can result in reduced hybridization efficiency. Ideally, hybridizations should be carried out in a fully enclosed air incubator in order to reduce evaporation from the sample. The hybridization ovens normally used for the screening of blots and libraries work well for this purpose. If an incubator is not available, a water bath or heating block may also be used. In the case of the heating block, it will be helpful to cover the tubes (e.g., with a thick layer of paper towels) to minimize condensation at the top of the tube.
18. We suggest starting with an RNase digestion buffer containing 3.75 μg/mL RNase A and 100 units/mL of RNase T1 (i.e., a 200X dilution of the stock RNase cocktail from Ambion). In theory, the relative and absolute amounts of RNase used in this assay are variable and may need to be optimized for each probe independently. In practice however, a very wide range of RNase concentrations generate effectively identical results *(9)* and thus, the suggested starting conditions are very likely to yield good protections. If you should desire to optimize

the RNase digestion conditions for your particular combination of probe and amount of target RNA, we suggest a range of dilutions varying from 1/100th to 1/1000th of the stock solution.
19. In the case of the undigested control, resuspend the pellet in approximately 20 µL of loading buffer and load 2 µL on the gel.

References

1. Melton, D. A., Krieg, P. A., Rebagliati, M. R., Maniatis, T., Zinn, K., and Green, M. R. (1984) Efficient in vitro synthesis of biologically active RNA and RNA hybridization probes from plasmids containing a bacteriophage SP6 promoter. *Nucleic Acids Res.* **12,** 7035–7056.
2. Lee, J. J. and Costlow, N. A. (1987) A molecular titration assay to measure transcript prevalence levels. *Methods Enzymol.* **152,** 633–648.
3. Goldrick, M. M., Kimball, G. R., Liu, Q., Martin, L. A., Sommer, S. S., and Tseng, J. Y. (1996) NIRCA: a rapid robust method for screening for unknown point mutations. *Biotechniques* **21,** 106–112.
4. Sagerstrom, C. G. and Size, H. (1996) RNA blot analysis, in *A Laboratory Guide to RNA* (Krieg, P. A., ed.), Wiley-Liss, New York.
5. Westerfield, M. (1993) *The Zebrafish Book*, University of Oregon Press, Eugene, OR
6. Sambrook, J., Frisch, E., and Maniatis, T. (1989) *Molecular Cloning*, Cold Spring Harbor Press, Cold Spring Harbor, NY.
7. Schenborn, E. T. and Mierendorf, R. C. (1985) A novel transcription activity of SP6 and T7 RNA polymerases: dependence on template structure. *Nucleic Acids Res.* **13,** 6223–6236.
8. Krieg, P. A. (1990) Improved synthesis of full-length RNA probes at reduced incubation temperatures. *Nucleic Acids Res.* **18,** 6463.
9. Krieg, P. A. (1991) Synthesis of RNA probes using SP6, T7 and T3 RNA polymerase. *Methods Gene Tech.* 1, 35–62.
10. Zinn, K., DiMaio, D., and Maniatis, T. (1983) Identification of two distinct regulatory regions adjacent to the human beta-interferon gene. *Cell* **34,** 865–879.
11. Newman, C. S., Grow, M. W., Cleaver, O. B., Chia, F., and Krieg, P. (1997) *Xbap*, a vertebrate gene related to bagpipe is expressed in craniofacial structures and in anterior gut muscle. *Dev. Biol.* **181,** 223–233.
12. Krieg, P. A., Varnum, S. M., Wormington, W. M., and Melton, D. A. (1989) The mRNA encoding elongation factor 1α, (EF-1α) is a major transcript at the midblastula transition in *Xenopus. Dev. Biol.* **133,** 93–100.
13. Tonissen, K. F. and Krieg, P. A. (1994) Analysis of a variant Max sequence expressed in *Xenopus. Oncogene* **9,** 33–38.

4

Quantitative Analysis of mRNA Levels in *Xenopus* Embryos by Reverse Transcriptase–Polymerase Chain Reaction (RT-PCR)

Oliver C. Steinbach and Ralph A. W. Rupp

1. Introduction

Over the last few years, RT-PCR *(1,2)* has become a widely accepted method for quantitation of steady-state mRNA levels, particularly in *Xenopus*. Its unmatched sensitivity and swiftness allows for a high sample throughput with minimal amounts of starting material—considerable advantages over the conventional methods of Northern blotting or RNase protection. Initially, the use of RT-PCR for quantitative analysis was viewed skeptically. This was based on the concern that minor differences in the reaction conditions between samples would erratically influence the exponential rate of PCR amplification; therefore, results would be skewed *a priori*. This theoretical concern has turned out to be irrelevant for most applications.

Here, we describe a basic protocol for RT-PCR analysis, which has been optimized in our laboratory. In the first step, total cellular RNA is reverse-transcribed into a random-primed, first-strand cDNA library. This library is then used as the template for PCR reactions, in which gene-specific primers amplify short regions from the respective cDNAs. The PCR products are trace labeled with radioisotopes during synthesis and thereby can be easily detected and quantitated after size fractionation by polyacrylamide gel electrophoresis. In its routine application, this method allows one to determine relative mRNA levels by using reference mRNAs for normalization. Ideally, test and reference cDNA templates are coamplified in the same tube ("multiplex" PCR). If required, absolute transcript numbers can also be obtained by a competitive PCR assay *(3)*. In general, our method requires less than 0.5 µg of total cellular

RNA per sample, measures reliably more than 100-fold differences in relative transcript abundance, and detects as little as 100 mRNA molecules; raw data from 50 or more samples can be obtained within 48 h.

Messenger RNA quantitation by RT-PCR strictly requires that PCR product amounts be directly proportional to the original mRNA template amounts. Establishing experimental conditions, which ensure a quantitative performance of RT-PCR, is usually thought to be a major obstacle. Many different parameters influence the performance of the assay, and conditions have to be established for each primer pair. Therefore, we have included in this chapter an elaborate guide for calibration of RT-PCR conditions, which has been evolved from practical experience. By no means does this guide claim to be exclusive, although in a few cases we felt obliged to give absolute values for critical parameters. We hope that it will help the reader save time and effort.

2. Materials
2.1. General Materials and Equipment

1. Automated thermal cycler.
2. Benchcoat.
3. Deionized (Millipore quality), autoclaved water (dH_2O).
4. Diethyl pyrocarbonate-treated H_2O (DEPC–H_2O). Caution: DEPC is extremely toxic.
5. 0.5 mL PCR tubes, 1.5- and 2-mL microfuge tubes.
6. Gloves.
7. Heating block (variable temperature).
8. Micropipet (two sets): 0.1–2 µL, 2–20 µL, 10–200 µL, 100–1000 µL, and appropriate tips.
9. Tabletop microcentrifuges with fixed-angle rotors at room temperature and 4°C.

2.2. RNA Purification

1. Chloroform (ACS grade), store at room temperature. Caution: chloroform is an irritant.
2. 70% Ethanol (ACS grade) in DEPC–H_2O, store at –20°C.
3. Isopropanol/tRNA: 60 µg/mL tRNA in 100% isopropanol (ACS grade), store at 4°C. Dilute the *Escherichia coli* tRNA (e.g., from strain MRE600, Roche Diagnostics, Mannheim, Germany, see **Note 1**) from a 10 mg/mL stock in DEPC–H_2O, store at –20°.
4. RNA isolation solvent for single step separation (e.g., TriStar™, AGS, Heidelberg). (Caution: The solvent contains phenol and is toxic.)

2.3. RT Reaction

1. RT–dNTP mix: 2.5 mM each dCTP, dGTP, dATP, dTTP diluted in DEPC–H_2O, store at –20°C.
2. 0.1 M dithiothreitol (DTT) in DEPC–H_2O, store at –20°C.

3. 10X RT buffer: 200 mM Tris–HCl, pH 8.4 (room temperature), 500 mM KCl, 30 mM MgCl$_2$, 1 mg/mL BSA in DEPC–H$_2$O, store at –20°C.
4. 100 µM RT-primer, deprotected, desalted, and redissolved in DEPC–H$_2$O: 5'-NNNNNC -3' (N = G, A, T, or C), store at –20°C.
5. 40 U/µL RNAsin™ (Promega, Madison, WI), store at –20°C.
6. 200 U/µL M-MLV reverse transcriptase, (Gibco BRL), store in aliquots at –70°C, sensitive to repeated freeze–thaw cycles).

2.4. PCR Reaction

1. 5 U/µL AmpliTaq™ DNA polymerase (Perkin Elmer, Branchburg, NJ), store at –20°C.
2. 370 MBq/mL (10 mCi/mL) [α-^{32}P]-deoxycytidintriphosphate, 110 TBq/mmol (3000 Ci/mmol) (Caution: radioactive material must be handled with great care and according to local radiation safety regulations!).
3. Light mineral oil.
4. PCR–dNTP mix (10 mM each: dCTP, dGTP, dATP, dTTP diluted in dH$_2$O), store at –20°C.
5. 10X PCR buffer: 100 mM Tris–HCl, pH 8.4 (room temperature), 500 mM KCl, 15 mM MgCl$_2$, 0.01 % (w/v) gelatine in dH$_2$O, made up from autoclaved stock solutions, store at –20°C.
6. 25 µM PCR primers, deprotected, desalted, and redissolved in dH$_2$O, store at –20°C.
7. Standard radiation safety devices.

2.5. Electrophoresis and Detection

1. Acrylamide/bisacrylamide solution (29:1) in dH$_2$O.
2. Blotting paper.
3. 5X DNA-loading dye: 0.25 % bromphenol blue, 0.25% xylene cyanol, 15% Ficoll 400 in dH$_2$O.
4. 10X TBE electrophoresis buffer: 1 M Tris–HCl, pH 8.6 (room temperature), 0.83 M boric acid, 10 mM EDTA.
5. Vertical polyacrylamide gel electrophoresis system with DC power supply.
6. Vacuum gel dryer (caution: use cooling trap or an activated charcoal filter to scavenge radioactive aerosols).
7. Phosphor imager system or, as minimal equipment, X-ray film and autoradiography cassettes with intensifying screens.

3. Methods

3.1. RNA Purification

Because of its speed, high sample throughput, and reliability, we favor a simplified version of the single-step extraction method by Chomczynski and Sacchi *(4)*. Stabilized, monophasic guanidinium thiocyanate–phenol extraction solutions are available from several manufacturers. Although RNA prepared by this method is contaminated with variable amounts of genomic DNA, this is usually not sufficient to generate false-positive signals (*see* **Note 3**).

1. Transfer embryos or tissue explants to autoclaved microfuge tubes (1.5 mL) and carefully remove excess buffer. Avoid shearing or air contact of specimen, in particular with tissue explants, as this may result in instant cell lysis and RNA degradation.
2. Add 1 mL per 50–100 mg tissue of Tristar™ solution, according to the manufacturer's instructions. For *Xenopus*, we use 200 µL per 3–5 embryos, or 100 µL per 3–5 tissue explants.
3. Add the same amount of Tristar™ solution to a separate tube, which is carried through the whole procedure as a mock RNA sample (*see* **Note 3**).
4. Vortex for 10 s, incubate for 5 min at room temperature and freeze at –70°C (although it is possible to proceed directly, freezing improves tissue lysis). Samples can be stored at this point for months, if required.
5. Thaw lysates at room temperature and vortex briefly.
6. Add 0.2 volumes of chloroform and vortex for 15 s.
7. Centrifuge samples (14,000g, room temperature, 5 min). The emulsion will separate into three phases. Transfer the upper, aqueous phase with the RNA to an autoclaved microfuge tube; avoid the interphase and the lower, organic phase, which contain genomic DNA and protein.
8. Add 0.5 starting volumes of isopropanol–tRNA solution. Mix by inverting the tubes several times.
9. Incubate for 15 min at room temperature.
10. Centrifuge (14000g, 4°C, 10 min).
11. Carefully remove the supernatant.
12. Add 500 µL cold 70 % EtOH and centrifuge (14,000g, room temperature, 5 min). Discard the supernatant, be careful to retain the RNA pellet.
13. Air dry the precipitated RNA for approximately 10 min at room temperature. Cover the open microfuge tubes (e.g., with lint-free paper wipes) to avoid airborne contamination.
14. Dissolve the RNA in DEPC–H_2O at a concentration of approximately 0.1–0.5 µg/µL. (One *Xenopus* embryo [prehatching stages] yields approximately 2–5 µg of total RNA; we add 25 µL DEPC–H_2O per embryo, or 2 µL DEPC–H_2O per tissue explant.)
15. Vortex and incubate for 10 min at 56°C. Collect the liquid by brief centrifugation. Store the RNA at –70°C.

3.2. RT Reaction

In this step, a single-stranded cDNA library is generated from total cellular RNA using short, random oligonucleotides and reverse transcriptase (RTase). Thus, each sample can be analyzed for the abundance of multiple mRNAs, including reference mRNAs, which is an important aspect for quantitative analysis (*see* **Subheading 3.4.**). We find that random hexamer oligonucleotides outperform oligo-dT as primers. Reverse transcription of some mRNAs (e.g., *Xenopus* MyoD [XMyoD]) is very temperature sensitive, presumably due to secondary structures. This problem is overcome by performing the RT reaction routinely at 55°C (the concommitant reduction of enzyme activity is irrelevant). To facilitate efficient primer hybridization, RT reactions are assembled on ice and started at 4°C. By using a precooled thermocycler, both the temperature ramp (60 s for our machines) and the incubation temperature are performed under reproducible conditions.

1. Prepare an RT premix (with a few extra reactions worth) on ice (the reaction may be downscaled twofold):

DEPC–H_2O	1.75 µL
10X PCR-HB buffer	1 µL
2.5 mM dNTP-Mix	2 µL
0.1 M DTT	1 µL
40 U/µL RNAsin™	0.25 µL
100 µM hexamer RT primer	1 µL
Total	7 µL

   Mix gently, centrifuge briefly, and store on ice.

2. Set aside the required amount of RT premix for minus RTase (–RTase) controls (*see* **Note 3**). Add 1 µL DEPC–H_2O per 7 µL premix, mix gently, centrifuge briefly and store on ice.
3. Thaw RTase at room temperature and place immediately on ice. Add 1 µL RTase per 7 µL of the remaining RT premix. Mix gently (do not vortex!), centrifuge briefly. Store plus RTase (+RTase) premix on ice.
4. Transfer 8-µL aliquots of the –RTase and +RTase RT premixes into 0.5-mL PCR tubes. Store on ice.
5. Add 2 µL of the RNA sample, mix gently, centrifuge briefly, and store on ice. Set up a sample, consisting of mock RNA (*see* **Subheading 3.1., item 3**, and **Note 3**) and (+RTase) RT premix.
6. Place tubes in a thermal cycler, precooled to 4°C, and incubate samples for 30 min at 55°C. Finally cool down to 4°C.
7. Centrifuge samples briefly and proceed with PCR amplification. Alternatively, RT samples can be stored at –20°C. After thawing and before use, spin down the RT samples.

3.3. PCR Reaction

In this step, RT samples serve as templates for PCR reactions, in which gene-specific primers amplify short regions from the respective cDNAs. It is preferable to coamplify several cDNA fragments in one tube ("multiplex" PCR), for instance, to obtain an internal standard for the quantitation of relative mRNA levels (*see* **Subheading 3.5.1.**), but this also saves considerable time and effort. Numbers and temperature profiles of PCR cycles have to be optimized for each primer pair and template (*see* **Subheading 3.4.**). After size fractionation by gel electrophoresis, specific PCR products can be identified by ethidium bromide staining. For accurate quantitation, however, we recommend radiolabeling the PCR products directly during synthesis by including trace amounts of [α-^{32}P]–dCTP in the PCR mix.

3.3.1. Standard Single Primer Pair or Multiplex PCR Reaction

1. Prepare a PCR mastermix (with a few extra reactions worth) on ice, which consists of the following components per reaction (the reaction may be downscaled twofold):

dH$_2$O	39.9 µL
10X PCR Buffer	4.8 µL
10 mM dNTP	1 µL
Primer mix—gene 1 (forward and reverse primer, 25 µM each)	2 µL
5 U/µL Taq polymerase	0.2 µL
10 µCi/µL [α-^{32}P]–dCTP (3000 Ci/mmol)	0.1 µL
Total	48 µL

   Additional primer pairs for multiplex PCR are accommodated by reducing the amount of dH$_2$O. Mix gently, centrifuge briefly, and store on ice.
2. Aliquot premix portions (48 µL or less, *see* **item 7**) into 0.5-mL PCR tubes.
3. Set up an extra reaction and add 2 µL dH$_2$O (the so-called minus template [–template] control; *see* **Note 3**).
4. Unless your thermal cycler is equipped with a heated lid, overlay each sample with two drops (approximately 50 µL) of light mineral oil.
5. Add 2 µL of RT reaction per PCR tube. Make sure you penetrate the mineral oil layer. Mix by pipeting up and down. Centrifuge briefly.
6. Place the reaction tubes in the thermal cycler and start the required PCR program.
7. Because of differences in template abundance, separate primer pairs may require different cycle numbers in multiplex PCR. In that case, primers for the more abundant mRNA are added to the PCR samples after $n-x$ cycle numbers (n is the cycle number for the less abundant mRNA; x: cycle number for the more abundant mRNA). PCR reactions are cooled to 4°C after $n-x$ cycles, and 2 µL of primer mix for each additional gene is added (final volume: 50 µL). Be sure to penetrate the mineral oil layer! Continue with the program for x cycles.
8. PCR products are stable and can be stored at room temperature or 4°C until further analysis.

3.3.2. Size Fractionation of PCR Products by PAGE

Although itself semiquantitative, polyacrylamide gel electrophoresis (PAGE) offers an inexpensive and convenient method for post-PCR analysis, in particular with large sample numbers. It is used to separate specific PCR fragments from primer–dimer or other nonspecific amplification products, as well as to remove unincorporated [α-^{32}P]–dCTP which would otherwise interfere with the quantitation of the specific PCR products.

1. Assemble and cast a standard 6% polyacrylamide gel with Tris-borate-EDTA (TBE)-buffer. Precast gels can be stored at 4°C for at least a week.
2. Assemble the gel in a vertical electrophoresis system. Prerun at 100 V for approximately 30 min (this is particularly necessary for gels that have been stored at 4°C).
3. Remove a 10-µL sample (or more) from the PCR tubes. Mix with 0.25 volumes of 5X loading buffer. If you find it difficult to pipette consistent amounts of PCR sample, remove the oil layer first.
4. Load samples with a flat tipped microsyringe needle or sequencing gel loading tips. Run the electrophoresis until the Bromphenol Blue dye front has reached the end of the gel. For our standard gel dimensions of 18 cm × 21 cm × 0.4 cm, this takes roughly between 3 (180 V initially, constant current) and 12 h (50 V initially, constant current).
5. Disassemble the glass plates and transfer the gel onto blotting paper. Label the orientation of the gel with a pen mark. Dry the gel under vacuum at 70–80°C for approximately 2 h.

3.3.3. Detection

The PCR products can be detected either by classical autoradiography on X-ray film with an intensifying screen at –70°C, or by storage phosphor technology *(5)*. During recent years, storage phosphor imaging has become the method of choice for quantitation, as it offers much higher sensitivity and dynamic range than X-ray films. If the latter are used, one should bear in mind that saturation levels have to be established by reference signals for each exposure. Irrespective of the detection method, PCR products should be visible within 12 to 24 h.

3.4. Calibration of RT/PCR Reactions

In theory, the amount of PCR product is directly proportional to the amount of starting DNA template and is exponentially proportional to the cycle numbers, according to the equation

$$N = N_0 (1+\text{eff})^n$$

Where N is the amount of DNA after n cycles, N_0 is the starting number of template DNA, eff is the amplification efficiency, and n is the number of cycles; *see* **ref. 6**. The sensitivity and the dynamic range of cDNA quantitation by PCR is therefore limited by the maximal amplification of a given template under nonsaturating conditions. This, in turn, depends on the template abundance, primer performance, and cycle number. In addition, the cDNA synthesis during the RT reaction must be representative, although, in practice, this is less of a problem.

3.4.1. Optimizing the RT Reaction

If the cDNA synthesis reaction is saturated by too much RNA, low abundance mRNAs may become underrepresented in the cDNA pool.

1. Add serial dilutions of an appropriate test RNA sample (*see* **Subheading 3.4.3.3.**) to the RT-reaction, e.g., in the range 0.1–2.0 µg.
2. Perform a standard PCR reaction with subsaturating cycle numbers.
3. Size-fractionate the PCR products on a polyacryamide gel.
4. Quantify the signals of specific PCR products. Plot the relative product amounts against the RNA input (*see* **Fig. 1A**). Doubling the RNA amount should lead to an increase in signal intensity by at least a factor of ≥ 1.8 (accommodating the average variability of the assay, *see* **Note 5**). In our experience, up to 1.0 µg of total cellular RNA can be added to the RT reaction without reaching saturation.

3.4.2. Design of PCR Primers

Here, we list some rules of thumb for the design of PCR primers. Additional information can be found in **refs. 7** and **8**.

1. Typical primers have a length of 20–30 nucleotides and a guanine/cytosine (G/C) content between 50% and 60%.
2. The last six bases should have a balanced G/C content to avoid mispriming in G/C-rich sequences.
3. Primers should not contain palindromic sequences or homopolymeric stretches, at least the last three bases must not be complementary to the same primer or to others used in the same PCR reaction.
4. The melting temperature of the primer–template hybrid should be between 60°C and 70°C. It should be similar for the forward (F) and reverse (R) primer ($\Delta T_m \leq 5°C$, if possible). In multiplex PCRs it is likewise important that the annealing temperatures of different primer pairs do not diverge too much. The T_m can be estimated for example by the equation $T_m = 69.3°C + (0.41 \times GC\%) - 650/\text{length}$.
5. The size of the amplified fragments should be in the range 100–600 basepairs. Whereas shorter products may comigrate with primer–dimer fragments, both the RT and PCR reactions select against longer fragments. In the case of multiplex PCR, specific products should differ in length by at least 50 bp to be separated during gel electrophoresis.
6. Primer pairs should hybridize to different exons (i.e., overspan an intron). This allows discrimination between mRNA-derived PCR products and potential false-positive signals amplified from genomic DNA (GD), which is present in the RNA preparations. If this is not the case, it may be necessary to verify results by (– RTase) RT reactions (*see* **Note 3**).

3.4.3. Selecting Suitable Templates for PCR Calibration

1. If available, use 1–10 pg linearized cDNA plasmid to test initially, whether primers will amplify the expected DNA fragment from the specific template. Such a reaction may be subsequently used as a size marker for the expected product. Diagnostic digests of plasmid- and mRNA-derived PCR products may be used to further verify the fragment identity.
2. Use 100 ng genomic DNA per reaction to test whether the primers overspan an intron (*see* **Subheading 3.4.2.**).
3. Use a standard cDNA sample prepared from a RNA template of embryos or tissues of a developmental stage, at which the gene of interest is maximally, or at least abundantly, expressed. Alternatively, a limited collection of cDNA libraries may be prepared from RNA samples of staged embryo populations (e.g., oocytes, blastula, gastrula, neurula). This collection of developmental reference templates can then be used each time new primers are tested.

3.4.4. Optimizing Annealing Temperatures and Cycle Numbers

There are no general PCR conditions that will work for every case. PCR conditions that have worked in many cases are given as a guideline from which to start:

Denaturation	30 s at 94°C
Annealing	30 s at 58°C
Extension	60 s at 72°C.

Perform 19 cycles for high abundance, ubiquitous transcripts (e.g., reference mRNAs; *see* **Note 4**) and 28 cycles for rare transcripts (e.g., from regulatory genes). These numbers accommodate the fact that the copy numbers of most transcripts will differ by less than three orders of magnitude (i.e., from 1–10 to 100–1000 copies per cell).

For quantitative RT-PCR, in particular multiplex PCR, the following three parameters are most important:

1. The annealing temperature must ensure a maximal difference in hybridization to specific versus nonspecific templates.
2. The amplification reaction must remain within the exponential phase (no plateau effect).
3. Primer pairs must not interfere with each other in multiplex PCR. Interference can result in the amplification of additional nonspecific products and/or reduction of absolute product amounts compared to single primer pair reactions.

Although these parameters can be tested separately, it is much more efficient to perform the tests in parallel. For this purpose we have developed a set of 11 diagnostic PCR reactions (*see* **Table 1**), which establishes conditions for two primer pairs simultaneously (e.g., to amplify a specific mRNA together with an internal standard).

Table 1
Diagnostic PCR Reactions for Primer Calibration

Reaction Type	Single Primer pair	Multiplex	Single Primers		−RTase control	−Template control	GD control
Primer pair 1	F/R	F/R	F	R	F/R	F/R	F/R
Primer pair 2		F/R		F R	F/R	F/R	F/R
Template	(+RTase) RT-reaction				−RTase RT reaction	H$_2$O	Genomic DNA
Cycles	x, x+2, x+4, x+6	x+6					

Primers: F—forward primer; R—reverse primer. Cycles: x depends on the respective template abundance and therefore has to be estimated for each primer pair (*see* above and **Subheading 3.3.1.**).

1. Set up three sets of PCR reactions, illustrated in table 1, each of which will be used to test a different annealing temperature (e.g., 7°C, 5°C, and 3°C below the calculated T_m). Use premixes wherever possible to minimize pipeting errors!
2. To monitor the product increase with cycle number, repeatedly remove 10 µL aliquots after x, $x+2$, and so on cycles. Size-fractionate PCR products and quantify their signal intensity. Use a labeled DNA molecular-weight marker (or a mixture of PCR fragments of known size) to identify specific products.
3. Plot relative product amounts against cycle numbers for each annealing temperature (*see* **Fig. 1A**). Determine the ratio in product amounts of single primer pair versus multiplex PCR reactions (*see* **Fig. 1B**).
4. Check the following criteria:
   a. The size of the amplified fragments must be correct.
   b. No bands must be visible in the single primers-control lanes.
   c. The GD-control lanes indicate whether primer pairs overspan introns or not. In the latter case, the −RTase-control lane may also show signals of the expected size, depending on the cycle number and the amount of genomic DNA contamination in the RNA sample (*see* **Note 3**).

Fig. 1. *(opposite page)* A multiplex RT-PCR calibration. The two mRNAs under investigation are XMyoDb and histone H4, the latter being used as internal standard (for primer sequences, *see* **refs. 12** and **18**), RNA from late gastrula embryos, when MyoD expression peaks, is used as the test template. The optimal temperature for the primers in the multiplex PCR was determined before (data not shown): 30 s 94°C, 30 s 58°C, and 1 min 72°C (eff ≥ 0.9). Samples were separated on 6% polyacryamide gels, dried, and exposed to a phosphor storage screen. Product amounts were quantified with a phosphor imager. The product amounts were calculated as arithmetic means from duplicate samples; for the sake of simplicity only one sample is shown. (**A**) Variation of cycle number and RNA amount. Note that additional nonspecific bands (∗) appear as more cycles are performed. For quantitation, the product amounts of the lowest cycle numbers, respective to RNA amount, were arbitrarily set at 1; the others

Analysis of mRNA Levels by RT-PCR

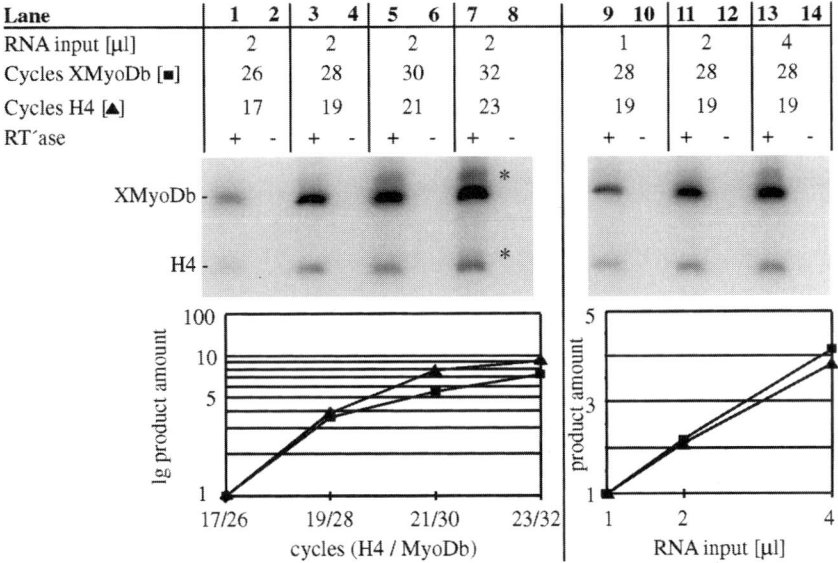

were calculated as multiples of this reference point. The following conditions were selected for routine analysis: 2 μL RNA input (i.e., 0.04 embryo or 1 animal cap equivalent) in each RT reaction, a total of 28 cycles for XMyoDb and 19 cycles for histone H4. **(B)** Amplification efficiency of primer pairs in single versus multiplex PCR. Additionally, single primer-, –RTase-, and -template-control PCR reactions were performed in parallel to exclude primer artifacts or contaminations. The ratio of product amounts in multiplex and corresponding single primer pair PCR is indicated. Genomic DNA-control lanes are not presented.

d. Compare the single-primer pair with multiplex-PCR reactions. The latter must not contain additional nonspecific products, and absolute product amounts should be similar.
   e. Determine for each annealing temperature the maximal cycle number, for which the amplification reaction remains within the exponential phase. If no plateau effect is observed, check whether the amplification efficiency (eff) is ≥ 0.75. If so, repeat the analysis with higher cycle numbers; if not, repeat the analysis with lower cycle numbers.
5. If one or more of these criteria are not fulfilled, test additional annealing temperatures. Try higher temperatures if nonspecific side products have been observed, and lower temperatures in case of low-amplification efficiencies (eff ≤ 0.75).
6. If this is not successful, design new primers which hybridize to a different region of the cDNA.

3.5. Quantitation

3.5.1. Relative mRNA Quantitation

An accurate quantitation of relative mRNA levels is best achieved by coamplifying a reference cDNA, which is used as an internal standard. Reference genes must not be influenced by the experimental conditions and, ideally, should be constitutively expressed at constant levels. mRNAs that are commonly used as reference templates in *Xenopus* and zebrafish are listed in **Note 4**. Most of these reference transcripts are very abundant and therefore require fewer amplification cycles than less abundant transcripts (for practical implications, *see* **Subheadings 3.3.1. and 3.4.4.**).

If test and reference cDNAs cannot be coamplified in the same tube, separate PCR reactions for both primer pairs should be analyzed with the same RT sample. However, one has to bear in mind that pipeting errors cannot be accounted for any more and that samples from different sets or different exposure times cannot be compared or normalized to each other.

1. Perform a multiplex PCR reaction with an internal standard. Alternatively, perform separate sets of PCR reactions for the gene(s) of interest and a reference gene. Size-fractionate the products on a polyacrylamide gel. Expose the dried gel and quantitate the signal intensity. Check that for the strongest signal on each gel the exposure is not saturated. If so, re-expose for a shorter time.
2. Quantify all of the PCR product signals for both the test and reference genes. Use background corrections (e.g., by integrating the signal intensity over the same area from adjacent positions above or below specific PCR products).
3. Determine the relative mRNA levels (P = product amount in arbitrary units):

$$\text{Relative mRNA level} = \frac{P_{\text{test gene}}}{P_{\text{internal standard}}}$$

4. These relative mRNA levels can be compared between different experimental conditions. By normalization to the relative mRNA levels of a reference sample (e.g., RNA from an untreated sibling control of the same developmental stage), x-fold changes in the relative mRNA amounts over the control are obtained.

3.5.2. Absolute mRNA Quantitation by Competitive PCR

The presented protocol is a variation of published methods *(3,6)* in which a specific competitor template is coamplified in the same tube by the same primer pair as the template of interest. We prefer to use synthetic transcripts as competitor rather than DNA because this also accounts for the relative efficiency of the RT reaction. Provided that the two templates (i.e., endogenous and competitor cDNAs) are amplified with very similar efficiency (*see* below), equivalent product amounts will be generated when equimolar concentrations of endogenous mRNA and competitor transcripts are present in the reaction. This point is empirically determined by adding serial dilutions of competitor transcript to a constant amount of cellular mRNA.

The competitor template is generated by introducing a unique restriction site into the amplified region of the cDNA of interest. As this will require only one- or two-point mutations, endogenous and mutant templates will later be reverse-transcribed and PCR amplified with near identical efficiency. While mRNA- and competitor-derived PCR products are initially of the same size, they are easily distinguished by cleaving the competitor-specific restriction site. An example for this application is given in **Fig. 2**.

1. Design a competitive cDNA template by introducing a unique endonuclease restriction site near the middle of the region, which is amplified by the PCR primers (for in vitro mutagenesis, *see* **ref. 9**). The selected restriction enzyme must be active in PCR-buffer.
2. Subclone the competitor DNA fragment into a suitable vector containing a T3, T7, or SP6 promotor and generate synthetic competitor transcripts (*see* Chapter 3). Determine the RNA concentration in a spectrophotometer. Knowing the concentration and molecular weight of the transcript, one can calculate the number of molecules per volume.
3. Prepare a RT premix, add proportional amounts of cellular mRNA, and set up a series of identical RT samples on ice. To these, add serial dilutions of competitor RNA. The dilution series should at least span one order of magnitude. Its concentration range depends on the relative abundance of the specific mRNA and has to be determined empirically. For an initial estimate, it helps to know the cell number corresponding to the mRNA amount tested. Medium abundant transcripts are present at about 10–100 copies per cell.
4. Carry out a standard PCR.
5. Digest competitor-derived PCR products with the appropriate restriction enzyme. Add 10–20 U restriction enzyme per PCR reaction and incubate for 30 min at the temperature required by the endonuclease.

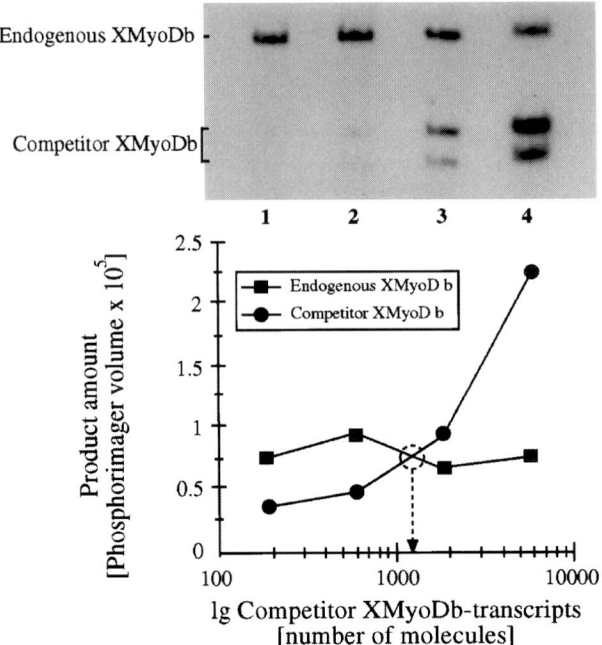

Fig. 2. Quantitation of absolute transcript numbers by competitive PCR. Serial dilutions of a mutated, synthetic XMyoDb transcript of known concentration are added to RT samples, each containing a constant amount of cellular RNA, equivalent to one animal cap explant from the blastula stage. By introducing two-point mutations, this mutant cDNA carries a HindIII restriction site near the middle of the amplified region, which distinguishes it from endogenous XmyoDb mRNA. After the PCR reaction, a 30-min restriction digest at 37°C is performed by adding 20 U HindIII to the PCR samples. PCR fragments derived from endogenous mRNA are not cut, but those derived from the competitor mRNA are cleaved into two smaller fragments (brackets). Samples were size-fractionated on a 6% polyacrylamide gel, dried, and exposed to a phosphor storage screen. Product amounts were quantified with a phosphor imager and plotted against the number of competitive template molecules. In this sample, 1.3×10^3 XMyoDb mRNA molecules were present per animal cap equivalent.

6. Size-fractionate PCR products and quantitate signal intensity as described. The signals of the two pieces of competitor-derived PCR products are added.
7. Plot radioactive product amounts of both the test and competitor against the number of competitive template molecules. Determine the number of mRNA molecules present in the test sample from the crossing point of the two graphs.

4. Notes

1. The trade names and suppliers of reagents that may be critical for quantitative assays are indicated. These may be substituted by other products, after checking their performance.

2. For safety precautions against DNA contamination we recommend wearing gloves at all stages, not handling cDNA plasmids or PCR products while setting up RT-PCR reactions, using a disposable benchcoat, setting aside boxes of pipet tips, microfuge and PCR tubes for exclusive use in RT-PCR; using separate sets of micropipets for (1) stock solutions, (2) RNA preparation, and (3) RT and PCR reactions, storing stock solutions in small aliquots, and avoiding aerosols by centrifuging microfuge tubes briefly before opening. Refer to **ref. 10** for additional measures, if necessary.
3. True-positive PCR products are identified by their length and a requirement for cDNA synthesis. False-positive signals arise from either exogenous DNA contamination (e.g., cDNA plasmids or PCR-product carry over) or from genomic DNA. The latter case is only of relevance if the primers do not overspan an intron and if the RNA preparations contain unusual amounts of genomic DNA. If necessary, determine the cycle number required to detect genomic DNA-derived PCR products from –RTase RT samples. With the method described here, this takes at least a 256-fold higher amplification (i.e., 8 cycles over maximal subsaturating cycle number). If the contamination with genomic DNA remains a problem, use signals from –RTase RT reactions as additional background correction. Alternatively, perform a DNAseI digest (for a protocol, *see* **ref. 11**) with each RNA sample.

    Generally, we perform three types of control reactions for each set of PCR samples. The first reaction template is a mock RNA sample (*see* **Subheading 3.1., step 3**), and the second with –RTase RT reaction(s) (*see* **Subheading 3.2., step 2**). The third type is a (-template) control (*see* **Subheading 3.3.1., step 3**). The appearance of PCR products in any of these samples indicates a DNA contamination of some sort and helps to identify the potential source. If this occurs replace potentially contaminated stock solutions and test again.
4. Commonly used reference mRNAs are histone H4 *(12)*, EF-1α *(13)*, orinithine-decarboxylase (ODC) *(14)*, GAPDH *(15)*, or fibroblast growth factor receptor *(16)* for *Xenopus*, and Max *(17)* for zebrafish.
5. We have consistently found that PCR-product amounts of independent, duplicate samples differ by less than 20% if the following rules are obeyed: Prepare samples of one experiment at a time; use premixes wherever possible; do not change stock solutions during a series. When preparing new stocks or using new lots of reagents, test these side by side against the previous ones. Check the performance of the thermo cyclers regularly using built-in test files and/or an external precision thermometer. Keep in mind that because of hardware differences, PCR conditions are thermo cycler-specific to some extent. This means, for instance, that PCR conditions obtained from the literature may need recalibration. Calibrate micropipets regularly—pipeting errors are the major source of variability!

References

1. Murakava, G. J., Wallace, B. R., Zaia, J. A., and Rossi, J. J. (1987) Method for amplification and detection of RNA sequences. European Patent Application 0272098.

2. Veres, G., Gibbs, R. A., Scherer, S. E., and Caskey, C. T. (1987) The molecular basis of the sparse fur mouse mutation. *Science* **237,** 415–417.
3. Gilliland, G., Perrin, S., and Bunn, H. F. (1990) Competitive PCR for quantitation of mRNA, in *PCR Protocols* (Innis, M. A., Gelfand, D. H., Sninsky, J. J., and White, T. J., eds.), Academic, New York, pp. 60–69.
4. Chomczynski, P. and Sacchi, N. (1987) Single-step method of RNA isolation by acid guanidinium thiocyanate–phenol–chloroform extraction. *Anal. Biochem.* **162,** 156–159.
5. Amemiya, Y. and Miyahara, J. (1988) Imaging plate illuminates many fields. *Nature* **336,** 89–90.
6. Wang, A. M., Doyle, M. V., and Mark, D. F. (1989) Quantitation of mRNA by the polymerase chain reaction. *Proc. Natl. Acad. Sci. USA* **86,** 9717–9721.
7. Kramer, M. F. and Coen, D. M. (1995) The polymerase chain reaction, in *Current Protocolls in Molecular Biology*, vol. 15 (Ausubel, F., M., Brent, R., Kingston, R. E., Moore, D. D., Seidman, J. G., Smith, J. A., et al., eds.), Wiley, New York, pp. 15.1.1–15.1.9.
8. Saiki, R. K. (1990) Amplification of genomic DNA, in *PCR Protocols* (Innis, M. A., Gelfand, D. H., Sninsky, J., J., and White, T. J., eds.), Academic, New York, pp. 13–20.
9. Higuchi, R. (1990) Recombinant PCR, in *PCR Protocols* (Innis, M. A., Gelfand, D. H., Sninsky, J. J., and White, T. J., eds.), Academic, New York, pp. 177–183.
10. Orrego, C. (1990) Organizing a laboratory for PCR work, in *PCR Protocols* (Innis, M. A., Gelfand, D. H., Sninsky, J. J., and White, T. J., eds.), Academic, New York, pp. 447–454.
11. Bauer, P., Rolfs, A., Regitz-Zagrosek, V., Hildebrand, A., and Fleck, E. (1997) Use of manganese in RT-PCR eliminates PCR artifacts resulting from DNAseI digestion. *BioTechniques* **22(6),** 1128–1132.
12. Niehrs, C., Steinbeisser, H., and De Robertis, E. M. (1994) Mesodermal patterning by a gradient of the vertebrate homeobox gene goosecoid. *Science* **263,** 817–820.
13. Krieg, P. A., Varnum, S. M., Wormington, W. M., and Melton, D. A. (1989) The mRNA encoding elongation factor 1-a (EF-1a) is a major transcript at the midblastula transition in Xenopus. *Dev. Biol.* **133,** 93–100.
14. Isaacs, H. V., Tannahill, D., and Slack, J. M. W. (1992) Expression of a novel FGF in the Xenopus embryo. A new candidate inducing factor for mesoderm formation and anteroposterior specification. *Development* **114,** 711–720.
15. Münsterberg, A. E., and Lassar, A. B. (1995) Combinatorial signals from the neural tube, floor plate and notochord induce myogenic bHLH gene expression in the somite. *Development* **121,** 651–660.
16. Lemaire, P. and Gurdon, J. B. (1994) A role for cytoplasmic determinants in mesoderm patterning: cell-autonomous activation of the goosecoid and Xwnt-8 genes along the dorsoventral axis of early Xenopus embryos. *Development* **120,** 1191–1199.
17. Schreiber-Agus, M., Horner, J., Torres, R., Chiu, F.-C., and Depinho, R. A. (1993) Zebra fish myc family and max genes differential expression and oncogenic activity throughout vertebrate evolution. *Mol. Cell. Biol.* **13(5),** 2765–2775.
18. Rupp, R. A. W. and Weintraub, H. (1991) Ubiquitous MyoD transcription at the midblastula transition precedes induction-dependent MyoD expression in presumptive mesoderm of X. laevis. *Cell* **65,** 927–937.

5

Wholemount *In Situ* Hybridization of *Xenopus* and Zebrafish Embryos

Joanne Broadbent and E. Mary Read

1. Introduction

Wholemount *in situ* hybridization (WISH) is a technique widely used to study the expression patterns of developmentally regulated genes. The last few years have seen massive improvements in the protocol. Not only can we now detect weak signals much more clearly but we can also visualize two, or even three, mRNAs in the same embryo. This allows a finer dissection of the spatial and temporal relationships between the expression of genes, even to the level of being able to show simultaneous expression of two genes within one cell.

The method uses RNA complementary to the endogenous mRNA ('antisense' RNA) which has been labeled with a particular antigenic label. We use digoxygenin- or fluorescein-labeled uridine-5'-triphosphate (UTP) in the production of our probes. These probes are hybridized to the embryo and visualized using anti-digoxygenin or anti-flourescein antibodies conjugated to alkaline phosphatase. Various chromogenic substrates for alkaline phosphatase are commercially available, and we describe the ones we routinely use. An increasing number of fluorescent substrates are becoming available, which should further improve the resolution of double *in situ* hybridizations. Fast Red (produced by several companies, but the one we use is from Boehringer Mannheim, UK) fluoresces under a rhodamine filter and can be used to visualize one probe, whilst the enzyme labeled fluorescence substrate ELF™ (Molecular Probes), which fluoresces under a DAPI filter, can be used for a second. For more detail about using fluorescent substrates, *see* **ref. *1***.

The following protocols for *in situ* hybridization are fairly widely used, but with some modifications of personal preference. For instance, for *Xenopus in situ* hybridization we tend to bleach our wild-type embryos because we have had some difficulty producing healthy albinos and these certainly will not withstand injection. In our hands, bleaching does not significantly reduce the signal. It is also particularly worth noting that both *Xenopus* and zebrafish *in situ* protocols benefit from the use of the Boehringer Mannheim blocking reagent in the antibody hybridization steps, greatly reducing nonspecific background and making the preabsorption of antibodies unnecessary.

The *Xenopus* wholemount *in situ* hybridization protocol presented here is based on that described by Harland *(2)*, but with some modifications. The zebrafish wholemount *in situ* hybridization protocol is based on the work of Jowett and Lettice *(3)*.

2. Materials
2.1. Probe Synthesis

1. Template cDNA in a vector such as pBluescript (Stratagene) or pGEM-7Zf (Promega) which contain T7 and T3, or T7 and SP6 promoters, respectively. To make antisense RNA, the template should be linearized at the 5' end of the cDNA, purified (*see* **Note 1**) and resuspended at 1 mg/mL in RNase-free water (*see* **Note 2**).
2. DIG-11-UTP NTP mix: 2.5 m*M* each ATP, cytidine-5'-triphosphate (CTP), GTP, 1.625 m*M* UTP, 0.875 m*M* DIG-11-UTP (Boehringer Mannheim) in water (or corresponding fluorescein-12-UTP mix, Boehringer Mannheim)
3. 5X Transcription buffer (Promega).
4. T7/T3/SP6 RNA polymerase (Promega).
5. RNase inhibitor (Promega).
6. 100 m*M* dithiothreitol (DTT) (Promega).

2.2. Fixing and Storing Xenopus *Embryos*

1. MEMFA: 0.1 *M* MOPS, pH 7.4, 2 m*M* EGTA, 1 m*M* MgSO$_4$, 3.7% formaldehyde (salts may be stored as a stock and the formaldehyde added immediately before use).
2. Methanol.
3. Sharp forceps or tungsten needles.
4. Sawn-off and flame-rounded Pasteur pipets.

2.3. Xenopus *Wholemount* In Situ *Hybridization*

1. PBSTw: PBS containing 0.1% Tween-20 (Sigma).
2. 25%, 50%, and 75% methanol in PBSTw.
3. Bleaching solution: 5% formamide, 0.5% SSC, 10% H$_2$O$_2$. **Extreme care is needed.** Make sure this solution is made just before use, as the reaction heats up and can explode if left for any length of time. Formamide should be diluted in water before addition of H$_2$O$_2$, otherwise an explosive mixture is formed.
4. 0.1 *M* triethanolamine.
5. Acetic anhydride.

6. WISH hybridization mix: 50% deionized formamide, 5X SSC, 1 mg/mL yeast RNA (Boehringer Mannheim) (*see* **Note 3**), 100 µg/mL heparin (Sigma), 1X Denhart's, 0.1% Tween-20 (Sigma), 5 mM EDTA. Store at 4°C.
7. 50% Formamide, 5X SSC; 25% formamide, 2X SSC; 12.5% formamide, 2X SSC.
8. 2X SSC, 0.1%Tween.
9. 0.2X SSC, 0.1%Tween.
10. Maleic acid buffer (MAB): 100 mM maleic acid, 150 mM NaCl, pH 7.5.
11. 2% Blocking reagent (Boehringer Mannheim) dissolved in MAB at 80°C and stored frozen.
12. Anti-digoxygenin or anti-flourescein F_{ab} antibody fragments conjugated to alkaline phosphatase (Boehringer Mannheim).
13. Alkaline phosphatase (AP) buffer: 0.1 M Tris, pH 9.0, 50 mM MgCl$_2$, 0.1 M NaCl, 0.1% Tween.
14. AP buffer + 10% PVA: 0.1 M Tris pH 9.5, 25 mM MgCl$_2$, 150 mM NaCl with 10% poly(vinyl alcohol) (PVA) 98–99% hydrolysed (31–50 kDa)(Aldrich) dissolved by boiling and cooling before use. Can be stored for a couple of weeks at 4°C but should be warmed (room temperature or 37°C) before use.

2.4. Fixation and Storage of Zebrafish Embryos

1. 4% Paraformaldehyde in PBS.
2. MS222 (3-aminobenzoic acid ethyl ester, methanesulfonate salt, Sigma) 0.03% solution in water.

2.5. Zebrafish Wholemount In Situ *Hybridisation*

1. PBSTw: PBS containing 0.1% Tween-20 (Sigma).
2. 25%, 50%, and 75% methanol in PBSTw.
3. 20mg/mL Proteinase K stock, stored at -20°C.
4. 2 mg/mL Glycine in PBSTw.
5. 4% Paraformaldehyde in PBS.
6. WISH hybridization mix: 50% deionized formamide, 5X SSC, 500 µg/mL yeast RNA (Boehringer Mannheim, see **Note 3**), 50 mg/mL heparin (Sigma), 0.1% Tween 20 (Sigma), brought to pH 6 using 1 M citric acid. Store at –20°C.
7. 25% Hybridization mix, 2X SSC.
8. 2X SSC/0.1%Tween, 0.2X SSC, 0.1% Tween-20.
9. 25% 0.2X SSC in PBSTw.
10. 50% 0.2X SSC in PBSTw.
11. 75% 0.2X SSC in PBSTw.
12. MAB: 100 mM maleic acid, 150 mM NaCl, pH 7.5.
13. 2% Blocking reagent (Boehringer Mannheim) dissolved in MAB at 80°C and stored frozen.
14. Anti-digoxygenin or anti-flourescein F_{ab} antibody fragments conjugated to alkaline phosphatase (Boehringer Mannheim).
15. AP buffer: 0.1 M Tris pH 9.0, 50 mM MgCl$_2$, 0.1 M NaCl, 0.1% Tween.
16. AP buffer + 10% PVA: 0.1 M Tris pH 9.5, 25 mM MgCl$_2$, 150 mM NaCl with 10% PVA 98–99% hydrolyzed (31–50 kDa)(Aldrich) dissolved by boiling and

cooled before use. Can be stored for a couple of weeks at 4°C but should be warmed (room temperature or 37°C) before use.
17. PBSTw, 20 mM EDTA.
18. 30%, 50%, and 70% glycerol in PBS.

2.6. Zebrafish Double Wholemount In Situ Hybridization

As for **Subheading 2.5.**, plus

1. 0.1 M glycine·HCl pH 2.2, 0.1% Tween-20.

3. Methods
3.1. Probe Synthesis

1. Template cDNA is linearised, purified (*see* **Note 1**) and resuspended at 1 mg/mL in water.
2. Set up 50 μL transcription reaction as follows:

5X Transcription buffer (Promega)	10 μL
100 mM DTT (Promega)	5 μL
DIG-11-UTP NTP mix	10 μL
Water	17.5 μL
RNasin (Promega)	0.5 μL (20 units)
DNA template (1 mg/mL)	2 μL
RNA polymerase T7/T3/SP6 (Promega)	5 μL (100 units)

   An optional ^{32}P–nucleotine-triphosphate (NTP) "spike" can also be added to determine the probe yield, but we routinely judge the amount and approximate length of RNA produced on an agarose gel.
3. Incubate for 2 h at 37°C.
4. Run 1 μL of the reaction on an agarose gel to check that transcript has been made. The single-stranded RNA should form a single band running at approximately twice the speed of the double-stranded DNA template and be clearly visible in just 1 μL of the reaction.
5. Add 10 units of RNase-free DNase I (Boehringer Mannheim) and incubate at 37°C for 15 min.
6. Add 1 μL 0.5 M EDTA to stop reaction and increase volume to 100 μL.
7. Take 1 μL for scintillation counting if "spike" was included.
8. Precipitate at –20°C for 10 min, adding 33 μL 10 M ammonium acetate and 250 μL ethanol.
9. Spin at 4°C for 15 mins, wash in 70% ethanol and resuspend in 100 μL water or hybridization mix.
10. Take another 1 μL for scintillation counting to calculate incorporation ratio if spike was added.
11. Probe can be stored at –20°C in either water or hybridization mix.
12. A titration using different dilutions of probe per hybridization reaction may be necessary to discover the optimum probe dilution, but we find diluting 1 μL of

probe stock into 200 μL hybridization mix is usually satisfactory and should result in a probe concentration of 200 ng/mL to 1 μg/mL.

3.2. Fixing and Storing Xenopus Embryos

1. Embryos are first devitellined with very sharp forceps or tungsten needles. Care should be taken not to damage the embryos but background staining is often reduced if the blastocoel is pierced at this stage.
2. Embryos are then transferred into MEMFA in glass scintillation vials and fixed for about 30 min. Longer than 1 h can affect signal intensity.
3. Replace the MEMFA twice with methanol and stand for another 30 min.
4. Replace this methanol with fresh methanol and store at −20°C.
5. Embryos can be stored like this for months but should be tested first if you are planning an important *in situ* and they have not been used for some time.

3.3. Xenopus Wholemount In Situ Hybridization

1. Embryos should be fixed and stored as described above.
2. Transfer embryos stored in methanol very carefully into 1.5-mL microfuge tubes using sawn-off Pasteur pipets.
   All subsequent treatments and washes are in 1 mL, unless stated otherwise.
3. Embryos should be rehydrated by washing in 75%, 50%, and 25% methanol in PBSTw and finally in PBSTw alone for 5 min each.
4. If using wild-type embryos, bleach to remove pigment by replacing PBSTw with bleaching solution and sitting them on a light box for 5–10 min. Invert occasionally to move the embryos, until the pigment can no longer be seen (*see* **Note 4**).
5. Bleaching solution should be removed by three, 5-min washes in PBSTw.
6. Embryos are then deacetylated by washing twice in 0.1 M triethanolamine, for 5 min each. To the second wash, add 2.5 μL acetic anhydride and a further 2.5 μL acetic anhydride after 5 min.
7. Triethanolamine and acetic anhydride are removed by two 5-min washes with PBSTw.
8. Replace the PBSTw with 0.5 mL hybridization buffer twice and prehybridize the embryos for at least 6 h at 60°C.
9. Following prehybridization, hybridization buffer should be replaced with fresh buffer containing 200 ng/mL–1 μg/mL nonradioactively labeled riboprobe (both probes should be added at this point for double labeling, *see* **Note 5**).
10. Hybridization is then allowed to proceed at 60°C, overnight, protected from light (*see* **Note 6**).
11. Following overnight incubation, hybridization buffer containing labeled probe should be removed and stored for further use (*see* **Note 7**). Replace with 50% deionized formamide/5X SSC and incubate for 10 min at 60°C.
    Following hybridization steps, pipet tips should be changed regularly to avoid contamination by probe carryover between samples.
12. Subsequent posthybridization washes, all at 60°C, are carried out with 25% deionized formamide/2X SSC for 10 min, 12.5% deionized formamide/2X SSC for 10

min, 2X SSC, 0.1% Tween for 10 min, and, finally, 0.2X SSC, 0.1% Tween for 30 min.
13. Rinse in PBSTw three times for 5 min each at room temperature.
14. Replace with MAB for 10 min and then block (ready for the antibody step) by incubation in MAB containing 2% Boehringer Mannheim blocking solution for 4–5 h. This blocking step may often be reduced to 2 h without additional background.
15. For antibody detection of probes, embryos are incubated with a 1:2000 dilution of anti-digoxygenin or anti-flourescein F_{ab} antibody fragments conjugated to alkaline phosphatase (Boehringer Mannheim) in MAB blocking solution overnight, rocking gently at 4°C. This step can be done for 4 h at room temperature if desired. (For double *in situ*, see **Subheading 3.4.**)
16. Excess, unbound antibody is then removed by washing at least five times, for an hour each, with MAB. These washes are carried out at room temperature, with gentle rocking.
17. For the alkaline phosphatase catalyzed color reaction, equilibrate with two washes in AP buffer, for 5 min each. After the second wash replace the AP buffer with either AP buffer containing 10% PVA and 3.5 µL BCIP (per milliliter of buffer) (Boehringer Mannheim) for turquoise color, or a 1 in 4 dilution of BM Purple (Boehringer Mannheim) for purple (*see* **Note 8**).
18. Transfer the embryos to dishes with a sawn-off Pasteur pipet and store in the dark at room temperature, observing the development of the color reaction periodically under a dissecting microscope. Color may be developed at 37°C, reducing developing time with no significant increase in background, or at 4°C.
19. Once the color reaction has developed satisfactorily, embryos should be washed in large volumes of PBSTw or AP buffer several times, especially if the reaction has been done in PVA-containing AP buffer (*see* **Note 9**), and then washed with methanol for at least 5 min to leach out low level, nonspecific background.
20. Next wash embryos in AP buffer again for 5 min and refix in MEMFA for at least 20 min.
21. Finally, dehydrate with methanol for 30 min at room temperature, replace with fresh methanol and store at –20°C.

3.4. Xenopus *Double Wholemount* In Situ *Hybridization*

For double-labeled *in situ* hybridization, two probes with different antigenic labels should be used. Digoxygenin and fluorescein are common choices. Both probes are hybridized at the same time but each antibody hybridization and color reaction are carried out separately.

The protocol is identical to that for Xenopus *single wholemount* in situ, *as detailed in* **Subheading 3.3.**, *up to the addition of antibody (step 15).*

1. Add the first antibody (*see* **Note 10**) diluted 1:2000 in MAB containing 2% Boehringer Mannheim Block. Incubate for 4 h at room temperature or overnight at 4°C with gentle shaking.

2. Wash five times in MAB for 1 h each, with gentle shaking.
3. Equilibrate with two washes in BM Purple buffer, for 5 min each. After the second wash replace the buffer with AP buffer containing 10% PVA and 3.5 µL BCIP (per milliliter of buffer) to develop the first probe in turquoise (*see* **Note 10**).
4. Allow color to develop in the dark as for single wholemount.
5. Wash embryos several times in PBSTw or AP buffer with gentle shaking to remove all PVA (*see* **Note 9**).
6. Equilibrate into methanol through a 25%, 50%, and 75% methanol series, 5 min each wash. To inactivate the alkaline phosphatase after the first color reaction is complete, embryos are stored in methanol for at least 1 h (we find that overnight is best if possible, although this may cause loss of signal for the second probe).
7. Wash embryos back into MAB through a 75%, 50%, 25% methanol series, 5 min each wash, and block in MAB plus 2% Boehringer Mannheim Block for at least 1 h prior to the addition of the second antibody.
8. Add second antibody diluted to 1:2000 in MAB containing 2% Boehringer Mannheim Block. Incubate for 4 h at room temperature or overnight at 4°C with gentle shaking.
9. Wash five times in MAB for 1 h each with gentle shaking.
10. Equilibrate with two washes in AP buffer for 5 min each. Replace with BM Purple diluted 1:4 in AP buffer and allow color to develop in dark.
11. Wash several times with PBSTw and equilibrate again into methanol. Wash with methanol for at least 5 min to leach out low-level, nonspecific background.
12. Fix in MEMFA for at least 20 min.
13. Finally dehydrate with methanol and store at −20°C.

3.5. Fixation and Storage of Zebrafish Embryos

1. Embryos are fixed in 4% paraformaldehyde in PBS overnight at 4°C or for 2 h at room temperature. They can safely be left for up to a week at 4°C without impairing the signal, however. Embryos younger than 15 somites are removed from their chorions after fixation; however, older embryos are first removed from their chorions using forceps and can be anesthetized in MS222 but must be washed several times in PBS before fixation.
2. Embryos are dehydrated through a 25%, 50%, 75% methanol in PBS series (5 min each wash) and stored in 100% methanol at −20°C. Embryos can be stored like this for many months.

3.6. Zebrafish Wholemount In Situ Hybridization

This protocol should give a clean signal, even with the weakest of probes. For very strong probes, however, a "quick" *in situ* protocol can be used, as described in **Note 11**.

All washes are done in a volume of 500 µL.

1. Rehydrate embryos through a 25%, 50% 75% methanol in PBS series and wash 4 × 5min in PBSTw.

2. Embryos older than 15 somites are treated with proteinase K (0.5 µL of 20 mg/mL stock per milliliter PBSTw) for 20 min at room temperature, *see* **Note 12**).
3. Stop proteinase K reaction by washing twice for 5 min in 2 mg/mL glycine in PBSTw.
4. Refix in 4% paraformaldehyde in PBS for 20 min at room temperature.
5. Wash 5 × 5 mins in PBSTw at room temperature with gentle shaking.
6. Briefly wash in 50% PBSTw, 50% hybridization mix (*see* **Note 13**) before prehybridizing in hybridization mix for at least 1 h at 65°C in inclined or horizontal heating block.
7. Replace prehybridization solution with fresh hybridization mix containing, typically, a 1:200 dilution of probe(s). For double *in situ* hybridization, both probes are added at this point (*see* **Note 5**). Hybridize overnight at 65°C in a heating block.
8. Remove probe in hybridization mix and store at –20°C for reuse (*see* **Note 7**).

    Following hybridization steps, pipet tips should be changed regularly to avoid contamination by probe carry over between samples.
9. The following washes are performed at 65°C for 10 min: 50% hybridization mix, 50% 2X SSC three times; 2X SSC, 0.1% Tween-20 once; 0.2X SSC, 0.1% Tween-20 four times.
10. The following washes are performed at room temperature for 5 min with gentle shaking: 75% 0.2X SSC, 25% PBSTw; 50% 0.2X SSC, 50% PBSTw; 25% 0.2X SSC, 75% PBSTw; 100% PBSTw. MABTw can be substituted for PBSTw in these washes.
11. Wash briefly in MAB before blocking with MAB containing 2% blocking reagent (Boehringer Mannheim) for at least 1 h at room temperature on a gently rocking table.
12. Replace block with anti-digoxygenin or anti-flourescein F_{ab} antibody fragments conjugated to alkaline phosphatase (Boehringer Mannheim) diluted in MAB containing 2% blocking reagent (Boehringer Mannheim). Anti-digoxygenin is diluted to 1:5000 and anti-fluorescein to 1:2000. Incubate for 2 h at room temperature or overnight at 4°C with gentle shaking. *(For double* in situ, *see* **Subheading 3.7.***)*
13. Remove antibody dilution and store at 4°C for reuse (*see* **Note 14**).
14. Wash eight times for 15 min each in either PBSTw or MABTw on a gently shaking table at room temperature.
15. Equilibrate embryos with AP buffer by washing three times for 5 min at room temperature.
16. For color reaction, transfer embryos to BM Purple (Boehringer Mannheim) diluted 1:4 in AP buffer for purple color, to AP buffer containing 10% PVA and 3.5 µL BCIP per milliliter for turquoise, or 0.1 *M* Tris–HCl pH 8.2 containing one Fast Red tablet (Boehringer Mannheim) per milliliter for red (*see* **Note 15**). Protect from light as color to develops. This can take from 10 min to several days, depending on the strength of the signal, but the reaction can be speeded up by incubation at 37°C. The reaction can be stored at 4°C overnight in AP buffer without significantly affecting the background.

17. Stop the reaction by washing in PBSTw containing 20 m*M* EDTA.
18. Nonspecific background from BM Purple or BCIP can be removed by washing in methanol at room temperature.
19. To permanently stop the reaction, refix in 4% paraformaldehyde for 15–30 min at room temperature.
20. Embryos can be stored in methanol at –20°C or in 70% glycerol (*see* **Note 16**) at 4°C. Take embryos through 30% and 50% glycerol in PBS before transferring to 70% glycerol. You may want to add azide to the glycerol for long-term storage.

3.7. Zebrafish Double Wholemount In Situ *Hybridization*

For double-labeled *in situ* hybridization, two probes with different antigenic labels should be used. Digoxygenin and fluorescein are common choices. Both probes are hybridized at the same time, but each antibody hybridization and color reaction are carried out separately.

The protocol is identical to that for zebrafish single in situ hybridization, as outlined in **Subheading 3.6.**, *until the addition of antibody* (***step 12***).

1. Add first antibody (*see* **Note 10**) diluted in MAB containing 2% Boehringer Mannheim Block. Anti-digoxigenin is diluted to 1:5000 and anti-fluorescein to 1:2000. Hybridize for 2 h at room temperature or overnight at 4°C with gentle shaking.
2. Wash in PBSTw or MABTw eight times for 15 mins shaking at room temperature.
3. First color reaction is performed in dark as above, using either BCIP for turquoise or Fast Red for red color (*see* **Note 15**).
4. Wash three times 5 min in PBSTw or MABTw shaking at room temperature.
5. To permanently stop the first color reaction, wash four times 5 min in 0.1 *M* glycine·HCl pH 2.2, 0.1% Tween-20 at room temperature.
6. Wash an additional four times in PBSTw or MABTw for 5 min each.
7. Return to 2% Boehringer Mannheim Block in MAB for at least 30 min before replacing with the second antibody diluted as above. Hybridize for 2 h at room temperature or overnight at 4°C, with gentle shaking.
8. Wash in PBSTw or MABTw eight times for 15 min each.
9. Equilibrate into AP buffer by washing three times for 5 min. For the second color reaction we routinely use BM Purple diluted 1:4 in AP buffer. Let color develop in dark.
10. Wash three times 5 min in PBSTw and fix in 4% paraformaldehyde to permanently stop the reaction.

4. Notes

1. Cut template can be purified by performing two phenol–chloroform–isoamylalcohol (24:25:1) extractions followed by two chloroform–isoamylalcohol (25:1) extractions.
2. We find it unnecessary to use DEPC-treated water and use autoclaved Milli-Q water for all RNase-free procedures.

3. Yeast RNA needs to be cleaned up by thorough phenol–chloroform–isoamylalcohol (24:25:1) extraction followed by chloroform–isoamylalcohol (25:1) extraction. Alternatively, MRE600 tRNA (BCL), which needs no purification, can be used.
4. Bleaching can be carried out at the end of the procedure after color development and some would argue that this gives a better signal intensity. However bleaching early allows you to see the color developing and is far better for doubles and signals that may be masked by the ectodermal pigment.
5. For double-labeled *in situ* hybridization two probes with different antigenic labels should be used (digoxygenin and fluorescein are popular). Both probes are hybridized at the same time, but each antibody hybridization and color reaction are carried out separately.
6. Fluorescein probes are light sensitive and better results are obtained if samples are kept in the dark after the addition of probe.
7. Used probes can be stored at –20°C and used up to three times.
8. 4.5 µL/mL NBT and 3.5 µL/mL BCIP (both Boehringer Mannheim) in AP buffer can be used for a purple color instead of BM Purple, but tends to give higher background.
9. PVA must be washed off with copious amounts of Tween-20 containing buffer before the embryos are exposed to methanol. Otherwise, a nonsoluble precipitate forms and sticks all over the embryos. However, it is very good at improving the signal using BCIP.
10. For double-labeled *in situ* hybridization, the first color to be developed was always the turquoise (BCIP alone) or red. This is due to the stability of the BM Purple, which cannot be washed out and thus disturbs the second color reaction. BCIP alone gives a nice turquoise color, however this is not a very strong reaction and should be used for the strongest signal. The antibodies for digoxygenin are much better than fluorescein and give a stronger signal; it is worth using this for the weakest signal. Try out all combinations of color and label type for the probes so that a very good resolution double can be obtained.
11. "Quick" fish *in situ* protocol is advisable for use only with strong probes:
    a. Fix embryos at room temperature for around 4 h in 4% paraformaldehyde in PBS.
    b. Wash 3 × 5 min in PBSTw.
    c. Prehybridize for 1 h at 65°C.
    d. Hybridize with probe overnight at 65°C.
    e. Wash in 2X SSC for 10 min.
    f. Wash in 0.2X SSC twice for 30 min.
    g. Wash in PBSTw twice for 5 min.
    h. Block for 1 h.
    i. Hybridize with antibody in Boehringer Mannheim Block for 2 h.
    j. Wash in PBSTw three times for 15 min.
    k. Equilibrate with AP buffer for 15 min.
    l. Proceed with the color reaction as above.
12. Proteinase K treatment makes early embryos too fragile and is unnecessary. For older embryos, longer than 20 min may be needed to increase probe penetration. Try 20 min per day of development.

13. Yeast RNA and heparin can be omitted from the hybridization mix used to make 50% PBSTw, 50% hybridization mix and 50% hybridization mix, 50% 2X SSC.
14. Antibody diluted in MABTw plus block can be stored at 4°C for reuse; however, storage for longer than a week is inadvisable.
15. Fast Red is soluble in methanol and turns the yolk of a zebrafish embryo yellow-orange. It is, however, fluorescent and, thus, extremely useful in double *in situ* hybridizations, where the expression domains of the two genes overlap. Use for the strongest signal, as it is weaker than BM Purple or BCIP.
16. Photographing zebrafish embryos is made easier by the viscosity of 70% glycerol.

References

1. Jowett, T. and Yan, Y.-L. (1996) Double flourescent *in situ* hybridisation to zebrafish embryos. *Trends Genet.* **12,** 387–388.
2. Harland R. M. (1991) *In situ* hybridisation: an improved whole-mount method for *Xenopus* embryos. *Methods Cell Biol.* **36,** 685–695.
3. Jowett, T. and Lettice, L. (1994) Whole mount in-situ hybridisations on zebrafish embryos using a mixture of digoxigenin-labelled and flourescein-labeled probes. *Trends Gents.* **10,** 73–74.

6

In Situ Hybridization to Sections of *Xenopus* Embryos

David Bertwistle

1. Introduction

Analysis of the spatial and temporal regulation of genes during embryogenesis is necessary if we are to understand their roles in developmental processes. *In situ* hybridization is the standard procedure used for describing the patterns of gene expression in embryos. In this technique, an antisense RNA probe, complementary to the mRNA from the gene of interest and labeled with radioisotope- or hapten-substituted ribonucleotides, is generated by an in vitro transcription reaction. The probes are hybridized to mRNA fixed in embryos, these are then washed under stringent conditions that only allow the probe to remain hybridized to the message. Finally, the probe is detected by autoradiography if it is radioactively labeled or by immunohistochemistry if it is hapten labeled. In the protocol described here, probes are labeled with the hapten-substituted ribonucleotide, digoxygenin-11-uridine-5'-triphosphate (DIG-11-UTP) and visualized by an anti-digoxygenin antibody conjugated to alkaline phosphatase, which catalyzes a chromogenic reaction.

In situ hybridization can be performed on wholemount or sectioned embryos. Wholemount embryos are much easier to use, as they require little preparation and many can be processed at once (*see* Chapter 5). For many purposes, wholemount *in situ* hybridization is sufficient. However, in some circumstances, it is preferable to perform *in situ* hybridization on embryo sections. One advantage of using sections is that probe penetration into the embryo is not a problem. This can be particularly important when analyzing gene expression in the yolky endoderm of *Xenopus* embryos *(1)*. In addition, although wholemount embryos can be sectioned following *in situ* hybridization, this can result in loss of signal intensity when using some detection systems and pre-

From: *Methods in Molecular Biology, Vol. 127: Molecular Methods in Developmental Biology: Xenopus and Zebrafish* Edited by: M. Guille © Humana Press Inc., Totowa, NJ

parative techniques. Finally, this technique can provide an alternative to performing wholemount *in situ* hybridization using two differently labeled probes. Co-expression of two differnt mRNAs in cells can be difficult to demonstrate using the techniques currently applied to *Xenopus*. Hybridization of the two probes to alternate sections of an embryo can often clarify whether a genuine overlap of expression exists *(2)*. This chapter describes a modification of the wholemount *in situ* hybridization protocol of Harland *(3)*, which can be used to detect specific mRNAs in sections of *Xenopus* embryos *(1,2)*.

2. Materials

2.1. Equipment

1. Microtome.
2. Dissecting microscope.
3. Hybridization oven.
4. Slide racks.
5. Coplin jars (screw-topped).
6. Wide-bore Pasteur pipet.
7. Watchmaker's forceps.
8. Scintillation vials.
9. Hot plate.

2.2. In vitro Transcription

1. Transcription buffer (Boehringer, Germany).
2. Dithiothreitol (DTT).
3. 2.5 mM DIG-NTP cocktail: 2.5 mM ATP, 2.5 mM CTP, 2.5 mM GTP, 1.67 mM UTP, and 0.83 mM digoxigenin-11-UTP (Boehringer, cat. no. 1 209 256). This cocktail can be stored at –20°C.
4. RNasin (Boehringer).
5. Sp6/T7 RNA polymerase (Boehringer).
6. DNase I, RNase-free (Boehringer).
7. TE (Tris–EDTA): 10 mM Tris-HCl (pH 7.5), 1 mM EDTA (pH 8.0).
8. Sodium dodecyl sulfate (SDS).
9. Sephadex G-50.
10. [α-^{32}P]–CTP (800 Ci/mmol) (Amersham, Buckinghamshire, UK).

2.3. Preparation of Sections

1. 2% Cysteine (pH 8.0).
2. MEMFA: 0.1 M MOPS (3-[N-morpholino] popanesulfonic acid) (pH 7.4), 2 mM EGTA (ethylene glycol-bis[β-aminoethyl ether] $N,N,N'N'$-tetraacetic acid), 1 mM MgSO$_4$, and 3.7% formaldehyde. A 10X solution of MOPS, EGTA and MgSO$_4$ can be stored at room temperature. Formaldehyde is added immediately prior to use.
3. Methanol.
4. Xylene.

In Situ *Hybridizations to* Xenopus *Embryos* 71

5. Embedding wax. 98% histoplast (Shandon, UK, cat. no. 67740006) and 2% beeswax (BDH, Poole, UK, 33018).
6. Poly-L-lysine-coated slides: Wash glass microscope slides in xylene, acetone and then absolute ethanol for 30 min each. Next rinse the slides in distilled water for 1 min and then soak them in a 50-μg/mL solution of poly-L-lysine in 10 mM Tris–HCl (pH 8.0) for 15 min. Finally, dry the treated slides at 40°C overnight. The poly-L-lysine used is >300,000 kDa (Sigma, Poole, UK, P1524).

2.4. In Situ *Hybridization*

1. Graded dilutions of ethanol in water (95%, 80%, 70%, 40%).
2. 4% paraformaldehyde in PBS: dissolve paraformaldehyde at 20% in distilled, autoclaved water, by heating to 65°C for 20 min and then adding NaOH until the pH is 8.0. Filter through filter paper. Prepare immediately before use.
3. SSPE: 20X stock is 3.6 M NaCl, 200 mM NaH_2PO_4/Na_2HPO_4 (pH 7.4), 20 mM EDTA.
4. Proteinase K (Boehringer).
5. 0.2 M HCl.
6. 0.1 M Triethanolamine, pH 8.0: add 1 mL acetic acid to 500 mL 0.1 M triethanolamine.
7. Acetic anhydride.
8. Denhardt's Reagent: 50X stock contains 5 g Ficoll (type 400, Pharmacia, UK), 5 g polyvinyl pyrrolidone, 5 g bovine serum albumin (BSA) and dH_2O to 500 mL.
9. Hybridization buffer: 50% deionized formamide, 5X SSC, 1X Denhardt's Reagent, 0.1% Tween-20, 0.1% CHAPS (Sigma C3023), 5 mM EDTA, 1 mg/mL, RNase-free tRNA (Boehringer), 100 μg/mL heparin.
10. Deionized formamide: Add 5 g amberlite resin (BDH) to 100 mL formamide. Stir for 30 min. Remove resin by pouring mix through filter paper. It is important to dispose of the filter paper with care, as formamide is a carcinogen and teratogen.
11. CHAPS (3-[3-cholamindopropyl)dimethylamino]-1-propanesulfonate) (Sigma, C3023).
12. RNase A (Boehringer): prepare by boiling for 10 min.

2.5. Digoxygenin Detection

1. PBT: phosphate-bufferd saline (PBS) with 2 mg/mL BSA and 0.1% TritonX-100 (Sigma, UK).
2. Heat-inactivated lamb's serum: Prepare by heating to 55°C for 30 min.
3. Anti-digoxigenin Fab fragments coupled to alkaline phosphatase (Boehringer, 1 093 274).
4. Color reaction substrates: 4-nitroblue tetrazolium chloride (NBT) (Boehringer, 1 383 213), 5-bromo-4-chloro-3-indolyl-phosphate (BCIP) (Boehringer, 1 383 221).
5. Color reaction buffer: 100 mM Tris-HCl (pH 9.5), 100 mM NaCl, 50 mM $MgCl_2$, 0.1% Tween-20, 1 mM levamisole (to inhibit endogenous phosphatase activity, add immediately before use).
6. 30% PBS, 70% glycerol.
7. Nail varnish.
8. Double-distilled and autoclaved water (*see* **Note 1**).

3. Methods
3.1. In Vitro Transcription of Digoxygenin-Labeled Riboprobes

The cDNA for the gene of interest must first be subcloned into a plasmid vector such as pGEM7 (Promega, Madison, WI) which has transcription initiation sites for SP6 and T7 RNA polymerases on either side of its multiple cloning site. Before in vitro transcription, the template plasmid should be linearized at a restriction enzyme site downstream of the desired, antisense transcript, to cause runoff of the transcript (*see* **Note 2**). It is advisable to also transcribe a sense probe, identical to the mRNA of interest, for use as a negative control.

1. Mix the following together at room temperature, in a 1.5-mL Eppendorf tube.

dH_2O	16.5 μL
5X transcription buffer	10 μL
100 mM DTT	5 μL
2.5 mM DIG-NTP cocktail	10 μL
DNA template (1 mg/mL)	2.5 μL
[α-^{32}P]–CTP (*see* **Note 3**)	0.5 μL
RNasin (20 units/μL)	0.5 μL
RNA polymerase (90 units)	5 μL
	50 μL

2. Incubate at 37°C for 2 h.
3. Add 5 μL of 1 mg/mL RNase-free DNase I and incubate at 37°C for a further 30 min to remove the DNA template.
4. Adjust the volume to 100 μL with TE + SDS such that the final SDS concentration is 0.1%. Remove 1 μL to allow determination of the total radioactivity in the transcription reaction.
5. Run the reaction through a 1-mL Sephadex G-50 column (equilibrated with TE + 0.1% SDS) to remove unincorporated nucleotides.
6. Transfer the 100 μL probe solution to a 1.5-mL Eppendorf tube, add 10 μL 3 M NaOAc and 250 μL ethanol and place the tube at –70°C for 10 min.
7. Spin at maximum speed in a bench microfuge at 4°C for 15 min and allow to dry on the bench (*see* **Note 4**).
8. Resuspend the pellet in 100 μL hybridization buffer. Take 1 μL for determination of radioactivity incorporated into the probe. Store the probe at –20°C. The probe is stable in this state for at least several months.
9. Quantitation. Count the Cerenkov radiation in the total reaction and in the RNA probe to determine the percentage incorporation. Total (100%) incorporation would correspond to 33 μg of probe.

3.2. Preparation of Tissue Sections

1. Dejelly *Xenopus* embryos in 2% cysteine (pH 7.8).
2. Remove the vitelline membranes of the embryos manually with a pair of watchmaker's forceps (Dumont, Switzerland).

3. Using a wide-bore Pasteur pipet, transfer the embryos to glass scintillation vials containing MEMFA fixative. Leave these to stand for 30 min to 1 h with occasional, very gentle agitation (*see* **Note 5**).
4. Remove half of the MEMFA and replace it with methanol. Then, being careful not to disturb the embryos, remove most of the MEMFA–methanol and replace with fresh methanol. Repeat and leave at room temperature for 30 min. Replace the methanol with fresh methanol and store the fixed embryos at –20°C in glass scintillation vials. Embryos can be stored in this way for several months.
5. To begin sectioning remove a vial of embryos from the freezer; once these are at room temperature remove most of the methanol and replace with absolute ethanol twice for 30 min, and then xylene twice for 30 min. Do not leave the embryos in xylene for any longer than this.
6. Replace the xylene with embedding wax at 60°C once for 30 min and once for 1 h.
7. Using a heated, wide-bore Pasteur pipet, transfer the embryos in molten wax to paper molds. Quickly arrange and orient embryos in the molten wax under a dissecting microscope with heated forceps (*see* **Note 6**). Leave these wax blocks at room temperature for several hours to harden.
8. Using a razor blade, carefully trim around the embryos so that they are left in a block of wax with a trapezium-shaped cross section raised above the rest of the block. Heat the main part of the block and stick it onto the chuck of the microtome.
9. Place the chuck in the microtome such that the blade hits the longest edge of the trapezium of wax first. Take 10- to 20-µm-thick sections.
10. Place a small amount of 40°C water onto a poly-L-lysine-coated slide. Carefully lay the sections onto the water. Remove any excess water and leave the slide on a 40°C hot plate for 2–3 d (*see* **Note 7**).

3.3. In Situ *Hybridization*

Steps 1–10 and **13** can be performed with the slides in a histology slide rack, using volumes of approximately 250 mL. However, for other steps it is advisable to use smaller volumes due to the cost of some reagents. **Steps 11**, **12**, and **14–16** are conducted using small volumes on horizontal slides, whereas **steps 17–21** are best done in Coplin jars. Screw-top Coplin jars are used to contain formamide fumes in **step 20**.

1. Dewax the sections with two 10-min treatments with xylene.
2. Rehydrate the sections by washing with a series of graded ethanols (absolute twice, then 95%, 80%, 70%, and 40%) for 1 minute each. From this point on, the sections must not be allowed to dry out.
3. Refix the sections with 4% paraformaldehyde in 1X PBS for 20 min.
4. Rinse in 2X SSPE.
5. Treat with proteinase K (at 3 µg/mL in 100 mM Tris–HCl [pH 7.5], 10 mM EDTA [pH 8.0]) at 37°C for 30 min.
6. Rinse in 2X SSPE.

7. Soak in 0.2 M HCl for 15 min.
8. Rinse in 2X SSPE.
9. Acetylate in 0.1 M triethanolamine (pH 8.0) with acetic anhydride. Add 1.25 mL acetic anhydride to 500 mL 0.1 M triethanolamine, stir for 5 min, then add another 1.25 mL acetic anhydride and stir for a further 5 min.
10. Rinse in 2X SSPE, and then in distilled H_2O.
11. Place small, broken cover slips on the slides, on either side of the sections, then add 300 µL hybridization buffer to each slide. Gently rest a whole cover slip on the broken cover slips, over the sections. The broken cover slip "shelf" enables easy removal of the cover slip in **step 12**. Incubate in a humid box at 60°C for 2 h to prehybridize the sections.
12. Carefully remove the cover slips and excess buffer from the slides and replace with 150 µL fresh hybridization buffer, containing the probe at 3 µg/mL. Place a fresh cover slip directly onto the sections, being careful not to trap any air bubbles. Hybridize in a humid box at 60°C overnight.
13. Transfer slides into 2X SSPE until the cover slips become dislodged.
14. Incubate the slides horizontally at 60°C in 1 mL per slide of hybridization buffer for 10 min.
15. Incubate the slides at 60°C in 1 mL per slide of 50% hybridization, 50% 2X SSPE + 0.3% CHAPS for 10 min.
16. Incubate the slides at room temperature in 1 mL per slide of 25% hybridization buffer, 75% 2X SSPE, 0.3% CHAPS for 10 min.
17. Transfer the slides to Coplin jars and incubate them at room temperature in 2X SSPE, 0.3% CHAPS for 30 min.
18. Wash the slides in 2X SSPE for 30 min.
19. Treat the slides with RNase A (at 20 µg/mL in 4X SSPE) at 37°C for 30 min.
20. Wash the slides in 50% formamide, 2X SSPE at 50°C for 1 h, changing the solution after 30 min.
21. Incubate the slides for 10 min in 2X SSPE, 0.3% CHAPS. Repeat.

3.4. Digoxygenin Detection

1. Rinse the slides for 2X 10 min in PBT.
2. Block for 1 h with PBT+ 20% heat-inactivated lamb serum.
3. Place the slides horizontally and incubate each with 1 mL of a 1:1000 dilution of anti-digoxigenin antibody Fab fragments coupled to alkaline phosphatase in PBT+ 20% lamb serum overnight at 4°C.
4. Wash for 5 × 30 min with PBT in a Coplin jar.
5. Wash in color reaction buffer for 5 min. Replace with fresh color reaction buffer containing NBT and BCIP (45 µL NBT and 35 µL BCIP per 10 mL color reaction buffer). Allow the reaction to proceed until the signal is at a sufficient intensity (*see* **Note 8**).
6. Once the signal is at the desired intensity, fix in MEMFA for 45 min.
7. Wash in PBT.
8. Place cover slips onto the slides, using a minimal amount of 30% PBS, 70% glycerol as the mountant.

9. Wipe the edges of the cover slips carefully with a tissue to remove excess mountant, then seal them onto the slides with nail varnish.

4. Notes

1. It is essential to avoid contamination by RNase before hybridization of the RNA probe to the section. Some protocols advise the use of DEPC-treated water to inactivate any contaminating RNase. However, DEPC is rather toxic, and as long as care is taken, distilled autoclaved water is quite sufficient.
2. Care should be taken in choosing the restriction enzyme for linearization as some RNA polymerases can initiate from 3' overhangs. This would produce a sense as well as antisense probe in the transcription reaction.
3. A small $[\alpha\text{-}^{32}P]$–CTP spike is included in the in vitro transcription reaction to allow quantitation of the RNA probe.
4. Probes of up to 1.3 kb have been used with no apparent problems with penetration into the sections. However, if much larger probes are used, it may be helpful to hydrolyze them. This can be done by resuspending the probe in 50 µL of 40 mM sodium bicarbonate–60 mM sodium carbonate and heating for 45 min at 60°C. This should hydrolyze the probe to a length of approximately 300 nucleotides. The hydrolyzed probe can then be precipitated by the addition of 200 µL dH$_2$O, 25 µL 3 M sodium acetate, and 600 µL ethanol.
5. Once the embryos have been fixed, they are much more fragile and must be treated with great care. This is crucial, as sections from damaged embryos fall off slides quite easily during the prehybridization treatments.
6. This is best done by transferring the embryos into preheated paper moulds, already containing molten wax. Sections are of a better quality if the embryos do not lie at the surface of the wax block. It is therefore advisable to let the wax at the very bottom of the mould cool and start to solidify before allowing the embryos to come to rest.
7. It is very important to allow the sections to dry properly and thereby stick to the slide well, as this appears to have a great impact on how well the sections survive *in situ* hybridization.
8. The color reaction can take from 30 min to 24 h to develop, depending on the level of expression of the gene of interest. It is important to monitor the color reaction on a negative control slide hybridized with a sense probe, in order to follow the development of background staining. If the reaction is left to proceed overnight, it is advisable to leave the slides at 4°C.

Acknowledgments

Many thanks to Emma Tiller and John Gurdon, who developed this protocol, for demonstrating it to me.

References

1. Lemaire, P. and Gurdon, J. B. (1994) A role for cytoplasmic determinants in mesoderm patterning: cell-autonomous activation of the *goosecoid* and *Xwnt-8*

genes along the dorsoventral axis of early *Xenopus* embryos. *Development* **120,** 1191–1199.
2. Walmsley, M. E., Guille, M. J., Bertwistle, D., Smith, J. C., Pizzey, J. A., and Patient, R. K. (1994) Negative control of *Xenopus* GATA-2 by activin and noggin with eventual expression in precursors of the ventral blood islands. *Development* **120,** 2519–2529
3. Harland, R. M. (1991) *In situ* hybridisation: an improved whole-mount method for Xenopus embryos. *Methods Cell Biol.* **36,** 685–695.

7

Zebrafish Immunohistochemistry

Rachel Macdonald

1. Introduction

Immunohistochemistry is a powerful technique for determining both the presence of and the subcellular location of proteins within tissues. Zebrafish are particularly amenable to this technique and it is possible to localize proteins both in whole embryos and larvae, as well as sectioned material (**Fig. 1**). In this chapter, I will describe the basic technique developed for wholemount labeling of zebrafish embryos and variations that are required in specific situations. In addition, for each technique, pitfalls will be highlighted and ways to avoid these suggested.

In zebrafish, immunohistochemistry is used to determine where and when proteins function. Double labeling with a cell-type-specific antibody and a new antibody allows the identification of which cell types and intracellular structures contain the protein of interest (e.g., **Fig. 1D**). For some antibodies, immunohistochemistry can also be used in conjunction with *in situ* hybridization of RNA to identify structures and cells expressing particular mRNAs *(3)*. Immunohistochemistry is also a useful technique for screening zebrafish lines carrying random mutations because of the ability to visualize subtle changes in the development of specific structures or cell types (e.g., neurons and axons).

The techniques described here are based on those outlined by Wilson et al. *(4)* and Patel et al. *(5)*. In summary, an antibody that has been raised in one species (e.g., mouse or rabbit) against the protein of interest is applied to fixed tissue which has been blocked for nonspecific antibody binding. After a period of incubation, usually overnight, the antibody is washed off thoroughly. A second antibody raised in a different species (e.g., goat) that recognizes the appropriate class of immunoglobulin for the primary antibody (e.g., mouse IgG) is then applied for a similar incubation time. The secondary antibody normally is

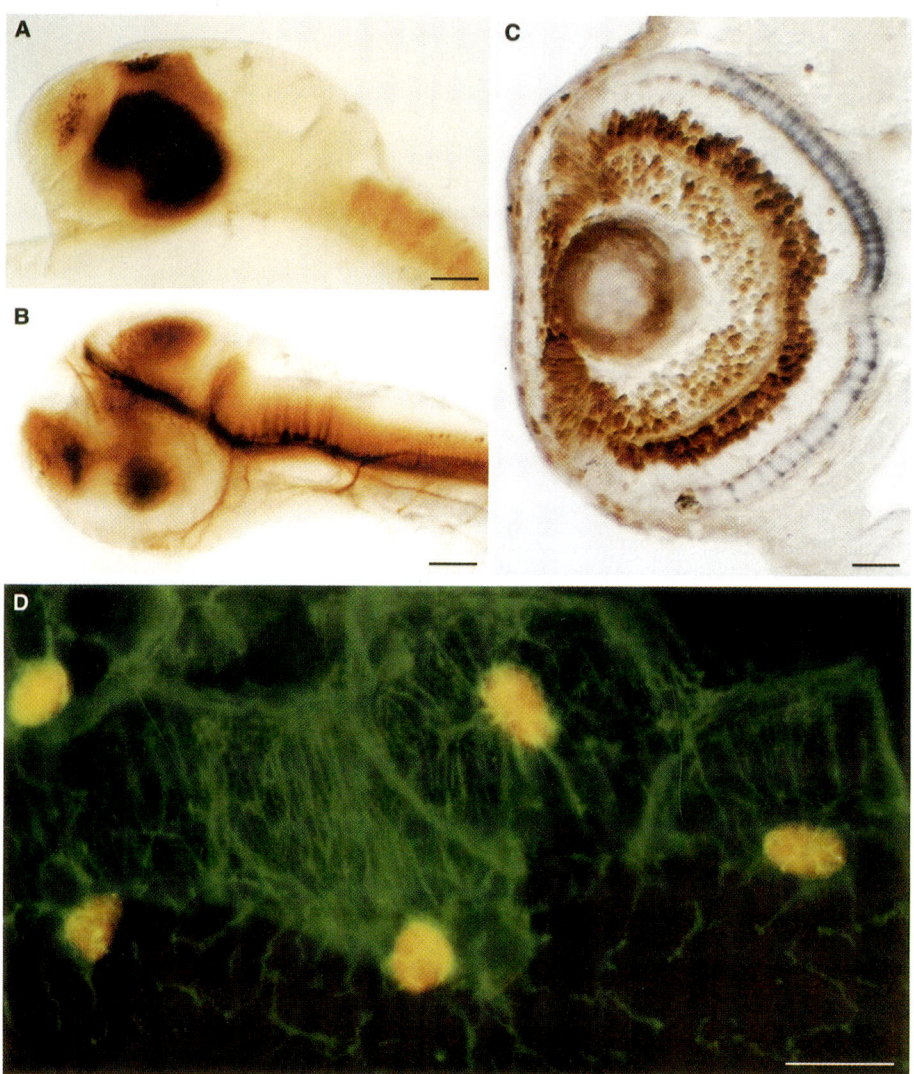

Fig. 1. Immunohistochemistry of wholemount and sectioned zebrafish tissue. (**A**) 28-h wholemount zebrafish embryo labeled with an antibody which recognizes the transcription factor, Pax6 (brown label) showing a lateral view with anterior to the left. (**B**) The 3-d larvae labeled with anti-acetylated tubulin antibody (brown), lateral view with anterior to the left. (**C**) Cross-section through a 3-d larvae eye labeled with anti-Pax6 antibody (brown) and a cone-specific antibody, fRet43 (black). (**D**) Oblique section through the surface of the optic nerve of an adult zebrafish labeled with anti-Pax2 antibody in yellow and anti-cytokeratin antibody in green and visualized using confocal microscopy. This preparation identified these Pax2 labelled cells as reticular astrocytes *(1)*. (**C** is reproduced from **ref. 2**; **D** is reproduced from **ref. 1**.)

conjugated to an enzyme such as horseradish peroxidase (HRP) or alkaline phosphatase (AP). This allows detection of the bound antibody complex by enzymatic reaction with a colored substrate such as diaminobenzidine (DAB). Although this technique is successful in most cases, there are some situations when other detection methods such fluorescence or amplification of the signal are required; these will be discussed in this chapter. Variations are possible at each step and the optimal conditions need to be determined for each particular antibody and application.

There are two major classes of primary antibodies *(6)*. Polyclonal antibodies are generally synthesized against peptides or large regions of proteins and may recognize a number of epitopes. This means that it is more likely that the antibody will cross-react between species and that successful labeling will be relatively insensitive to fixation. However, nonspecific background may also increase. Monoclonal antibodies, synthesized using mouse hybridomas, recognize single epitopes and, thus, generally have lower backgrounds. This also means that unless the protein is highly conserved, the antibody is less likely to recognize the epitope in other species and may be sensitive to fixation.

The method is divided into four main sections. First, a general protocol will be outlined for wholemount labeling of zebrafish embryos (*see* **Figs. 1A,B**). The second section covers variations to this protocol for larvae wholemounts to overcome the difficulties of antibody penetration in older tissue. This method, based on protocols established for medaka larvae *(7,8)*, was developed for use specifically with anti-acetylated tubulin antibody on 2- to 6-d-old larvae (*see* **Fig. 1B**) and may not apply to other primary antibodies. Labeling of tissue from larvae older than 1 wk and from adults is often difficult using the wholemount technique, as penetration of the antibody is likely to be incomplete. To overcome this problem it is necessary to section the material prior to labeling (**Figs. 1C,D**); this is covered in the third section. The final section covers the technique for labeling the same tissue with two different antibodies, and mounting and sectioning labeled tissue will be outlined. Discussion of the types of primary antibodies available, the production of antibodies and the choice of antibody will not be covered, as this is beyond the scope of this chapter.

2. Materials
2.1. Wholemount Protocols

1. Immunohistochemistry requires almost no special equipment. For experiments on a small number of different samples, all incubations, except for the final detection step, may be carried out in 1.5-mL microcentrifuge tubes. Limit the number of embryos in a tube to a maximum of 50 and use a volume of 500 µL for washes and 300 µL for antibody incubations. To minimize the labor involved for

larger experiments, use embryo holders that sit in multiwell plates as described in *The Zebrafish Book (9)*. As the detection step must be monitored under low magnification, it is most convenient to transfer the embryos to 24-well plates for this step. A rotating table is required to mix embryos in solutions during most incubation steps and it is useful to have another at 4°C for antibody incubations.

2. PTU (0.2 mM phenylthiocarbamide; Sigma, Poole, UK): make a 10X stock (dissolve 30 mg in 100 mL of embryo-rearing medium; **ref. 9**) and store at 4°C, dilute when required. *Caution*: PTU is extremely toxic and must be weighed in a fume cupboard and protective clothing worn.

3. Anaesthetic—0.03% MS222 (3-amino benzoic acid ethyl ester; Sigma): Make a 10X stock (dissolve 300 mg of powder in 100 mL rearing medium; **ref. 9**) and store at 4°C, dilute when required. *Caution*: This compound is a potent carcinogen; thus, precautions must be taken when handling.

4. 4% paraformaldehyde (PFA), pH 7.4: Mix 4 g of paraformaldehyde in 50 mL of distilled water (dH$_2$O) and heat to 60°C in a fume cupboard. Generally, the powder will dissolve with vigorous stirring with the temperature maintained at 60°C for several minutes, but a few drops of 1 N NaOH may facilitate the process. The solution is cooled and 50 mL of 2X phosphate-buffered saline (PBS) added and the pH determined. Store frozen as aliquots or at 4°C for up to 2 d.

   PIPES-bufferd formadehyde (PEM) fix: 9 parts PIPES (0.1 M PIPES, 2 mM EGTA, 1 mM MgSO$_4$, pH 6.95 with HCl) and 1 part 37% formaldehyde. Store at 4°C.

   Bouins fix: 2 g picric acid dissolved in 500 mL dH$_2$O, filter through Whatman number 1, add 20 g paraformaldehyde, heat to 60°C in a fume hood and add a few drops of 1 N NaOH to dissolve. Cool and add 500 mL 2X PBS.

   2% Trichloroacetic acid (TCA): make up a 10% TCA stock diluted in dH$_2$O, dilute to 2% using PBS.

5. PBS: 8. g NaCl, 0.2 g KCl, 1.44 g Na$_2$HPO$_4$, 0.24 g KH$_2$PO$_4$ in 1 L dH$_2$O, pH 7.4, can be made from commercially available tablets or powder.

   An alternative to PBS is 0.1 M PB, pH 7.4 made from two stock solutions: 38 mL of A (0.4 M NaH$_2$PO$_4$·2H$_2$O–31.2 g/500 mL dH$_2$O), 162 mL of B (0.4 M Na$_2$HPO$_4$–28.4 g/500 mL dH$_2$O) and 600 mL of dH$_2$O.

   Phosphate-buffer plus triton (PBT): PBS or 0.1 M PB plus 0.8% Triton X-100 (Sigma).

6. Trypsin: 2.5% trypsin stock is purchased commercially (e.g., Gibco, Paisley, UK) and stored at –20°C. Dilute to 0.25% with PBS prior to use.

7. Goat serum and other nonimmune sera: heat-inactivated stock is available commercially (Sigma), store in aliquots at –20°C.

8. Store antibodies in concentrated form in small aliquots at –20°C. When required, thaw an aliquot and store at 4°C while in use. Avoid freeze–thawing any antibody as much as possible. For dilute monoclonal antibody supernatants, store aliquots at 4°C. Dilute antibodies according to the suppliers information; however, it is often useful to determine the optimal concentration by carrying out a series of dilutions.

9. DAB: Store aliquots of stock (500 mg/10 mL dH$_2$O) at –20°C and thaw prior to use. Dilute 500 µL stock in 30 mL PBS and use within 30 min. *Caution*: DAB

is a potential carcinogen, so precautions must be taken when handling (*see* **Note 5**).

To achieve a black reaction product add 500 µL each of nickel sulfate solution (1% [NH_4]$_2$$SO_4$·$NiSO_4$·$6H_2O$) and cobalt chloride solution (1% $CoCl_2$) to 30 mL of DAB solution dropwise while stirring to prevent precipitation.

10. Mounting glycerols: 30%, 50%, and 70% glycerol in PBS.

2.2. Sectioning Protocol

1. Cryostat sectioning requires a supply of liquid nitrogen. For labeling sections, use a humidified box made by attaching two pipets parallel to each other on the bottom of an airtight box separated by less than the length of a microscope slide. This allows buffer to be layered into the box with the slides suspended above.
2. Agarose–sucrose: 1.5% agarose, 5% sucrose in PBS, heat to dissolve, and aliquot. These can be stored at −20°C.
3. Sucrose: 30% sucrose in PBS, filter sterilize, and store at 4°C in aliquots.
4. OCT mounting medium is available commercially (BDH [Poole, UK], Agar Scientific [Stansted, UK]).
5. TESPA (3-aminopropyltriethoxysilane; Sigma) coated slides: Dip slides in 2% TESPA in acetone for 30s, rinse twice in 100% acetone and air dry.
6. DAB: dissolve a 125-µL aliquot of DAB in 7.5 mL of PBS, and, when ready to use, add 40 µL 6% H_2O_2.
7. Toluidine Blue: 0.02% Toluidine Blue in dH_2O, filter through Whatman No. 1 regularly to remove precipitate.

3. Methods

3.1. Wholemount Immunohistochemistry of 0- to 40-h Zebrafish Embryos

1. Fixation: The method chosen depends on the sensitivity of the antigen to the fixation process, as well as the preservation and permeability of the tissue after fixation, and needs to be determined for each antibody (*see* **Note 1**). The most commonly used fixative for embryos at this age is a solution of 4% paraformaldehyde or other formaldehyde solutions such as the PEM fix. Fixation is normally carried out for 3 h at room temperature or overnight at 4°C, although fixation times may differ for each antigen. Embryos of less than 18 somites can be left in their chorions during fixation and dechorionated during washing steps. As zebrafish embryos are small, fixation of 20–50 embryos can be carried out in 0.5–1.0 mL volume of fixative in microcentrifuge tubes. After fixation is complete, replace the solution with an equivalent volume of phosphate buffer (either PBS or 0.1 *M* PB). It is possible to store the embryos in PBS with residual fixative for several weeks at 4°C before antibody labeling. Prior to labeling, wash the embryos 5 × 5 min with PBS to remove all traces of fixative.
2. Blocking: Before application of the primary antibody, nonspecific binding sites for immunoglobulins and other types of background protein interactions must be

blocked. This requires incubation of the embryos in a solution of serum, preferably from the same species as the antibody (e.g., goat serum). Dilute the serum to 10% (v/v) in PBT and incubate 20–30 embryos in 0.5 mL of this solution for 1 h at room temperature on a rocking table.
3. Primary antibody incubation: Dilute the primary antibody in PBT plus 1% goat serum to the appropriate concentration, as recommended by the supplier. Remove the blocking solution with a pipet and apply the primary antibody in a minimum volume of 100 µL (depending on the number of embryos). Either incubate the embryos overnight at 4°C on a rocking table with gentle agitation or for shorter periods at room temperature. The length of incubation is flexible and it is possible to leave embryos in primary antibody for several days. In some cases, longer incubations may improve antibody penetration. When testing new antibodies, it is necessary to include a number of controls outlined in **Note 2**.
4. Washing: after the incubation is complete, remove as much of the primary antibody as possible and replace with a large volume of washing solution (PBT). Where the primary antibody is in short supply, reuse the diluted solution several times, as normally the antibody is present in excess. Change the wash three to five times, with at least 15 min for each wash with gentle rocking at room temperature. The actual length of wash is not critical; however, generally, the background will be reduced further with longer washes over several hours (**Note 3**).
5. For embryos older than 30 h, it is necessary to quench endogenous peroxidases if the HR–DAB system described here is used. Wash the embryos for 5 min in 50% methanol–PBS solution, and finally in 100% methanol. Replace with the inactivating solution (50 µL 6% H_2O_2 in 1 mL 100% methanol) and incubate for 10 min at room temperature. Wash for 5 min with 50% methanol–PBS solution, PBS, and, finally, several times with PBT.
6. Dilute the secondary antibody (*see* **Note 4**) as recommended by the supplier in PBT + 1% goat serum and apply to the washed embryos. Incubation is similar to that of the primary antibody, either overnight at 4°C or several hours at room temperature. If secondary antibodies conjugated to a fluorescent chromogen are used, keep the embryos in the dark from this step on to prevent bleaching.
7. Remove the antibody and wash the embryos in PBT as described in **step 4**. Finally transfer the embryos to PBS alone prior to the detection steps. At this point, if fluorescence is used, clear the embryos in graded glycerols and mount in 70% glycerol containing an antibleaching agent such as Citifluor (Agar Scientific Ltd).
8. Dilute the DAB stock immediately before use. Care must be taken with this reagent; *see* **Note 5** for tips on handling and disposing of DAB solution and plasticware that has been in contact with DAB. For a black reaction product, add the nickel and cobalt solutions to the DAB. Transfer the embryos to multiwell plates, remove excess PBS, and replace with DAB solution. Preincubate for 20 min; then add 1–2 µL of 6% H_2O_2 per well and monitor the reaction at low magnification under a microscope. The reaction should be complete in 10–20 min (**Note 5**). To stop, carefully transfer the embryos back to microcentrifuge tubes containing PBS and wash several times with PBS, collecting the waste in bleach

(**Note 5**). Clear the embryos through graded glycerols over several hours, allowing them to sink to the bottom of the tube before changing the solution. Dissect, mount, and store the embryos in 70% glycerol. Alternative clearing methods include dehydrating rapidly in graded ethanols (70%, 90%, 95%, twice in 100%; 5 min each), transferring to pure methyl salicylate solution and mounting in Permount or DPX. Embryos cannot be stored in methyl salicylate indefinitely, as eventually the staining will be cleared.

9. If the labeling is weak, amplification can be carried out by using a three-tier detection method such as the biotin–strepavidin system (**Note 6**). Application of a permeabilization step after fixation and before blocking (as described in **Note 7**) can also increase penetration and labeling.

3.2. Wholemount Antibody Labeling of Zebrafish Larvae (2–6+ d)

1. Rear larvae in PTU diluted in rearing medium to prevent pigment formation *(10)*. At the required stage, wash larvae in rearing medium without PTU, anesthetize in 0.03% MS222, wash in PBS briefly, then fix in 2% TCA for 3 h at room temperature. Fixation should not be longer than 4 h. After fixation, wash the tissue in PBS for 3 × 5 min, and store at this stage at 4°C until required. Care should be taken when handling the tissue during this labeling technique, as the larvae are fragile after fixation.
2. Prior to permeabilization, wash tissue twice in PBT at room temperature, then chill on ice. Replace the PBT with 200–300 µL of chilled trypsin solution (0.25% in PBT) and incubate on ice using the following times as guidelines:

    2–3 d old: 4 min
    3–4 d old: 5 min
    5–6+ d old: 6 min

    Keep all solutions cold and take care not to extend incubation times. Replace the trypsin solution immediately with PBT and wash 5 × 5 min at room temperature. For alternative permeabilization methods, see **Note 7**.
3. Block nonspecific binding sites with 10% goat serum as described in **Subheading 3.1., item 2**.
4. Replace with diluted primary antibody (as described in **Subheading 3.1., item 3**) and incubate at least overnight at 4°C on a rocking table. With older stages, better results can be obtained by incubating for longer periods up to several days and this should be determined for each antibody.
5. Remove the antibody and wash over several hours with PBT, changing the buffer at least five times. Finally, wash with PBS.
6. It is necessary to quench endogenous peroxidases in larvae as described in **Subheading 3.1., item 5**.
7. Incubate in secondary antibody diluted as described in **Subheading 3.1., item 6**, generally overnight or longer at 4°C on a rocking table.
8. Wash the larvae over several hours in PBT, changing the wash at least five times. Finally wash the larvae in PBS.

9. The detection step is carried out in the same manner as described in **Subheading 3.1., item 8**. As TCA-fixed tissue is fragile, it is necessary to postfix in 4% PFA at 4°C overnight, as this makes it easier to handle the labeled tissue. The tissue is washed in PBS and cleared to 70% glycerol for dissection and mounting.

3.3. Labeling Sectioned Material

The best method of sectioning zebrafish, in terms of preservation of tissue morphology and antigenicity, is to use cryosectioning techniques (**Note 8**).

1. Fix tissue in the appropriate fixative as described for wholemount procedures (*see* **Subheading 3.1.**). Generally, for larger pieces of tissue such as adult brains, a longer fixation period is required, such as 48 h at 4°C, then wash the tissue several times in PBS prior to embedding for sectioning.
2. Prepare the tissue for sectioning as described in *The Zebrafish Book* (**9**). Remove the tissue from the wash, place in a small mold, and drain the excess PBS with filter paper. Fill the mold with molten agarose–sucrose solution, ensuring that the temperature of the agarose is less than 50°C. As the agarose solidifies, orient the specimen as required. When the agarose is completely solid, remove the block from the mold and trim to a flat-topped pyramid shape containing the specimen close to the top cutting surface, ensuring that the base is wider than the top and the pyramid is a squat shape. Place the agarose block in a large volume of 30% sucrose solution (2–5 mL, depending on the size of the block) and store at 4°C at least overnight until the block has sunk to the bottom of the tube. Blocks may be stored this way for several weeks prior to cutting.
3. Cool the cryostat to –30°C. Freeze platforms of OCT solution or similar commercial cryo-embedding compounds onto the cryostat specimen chucks. Remove the agarose specimen block from the sucrose solution, drain off the excess sucrose on filter paper (e.g., Whatman No. 1), and carefully mount onto the frozen OCT platform using a small amount of fresh OCT. Freeze the specimen onto the cryostat chuck by immersing most of the chuck into liquid nitrogen. Avoid immersing the whole specimen under liquid nitrogen, as the OCT can fracture at very low temperatures. Place the chuck in the cryostat chamber and allow to equilibrate to –30°C. Check that the specimen is still attached firmly to the chuck and reattach with water if loose. Cut 7- to 10-µm sections, collect on TESPA-coated slides (**Note 9**), and air-dry for several hours to ensure that the sections stick to the glass. Slides can be stored for several months, wrapped in foil at –20°C. When required, remove from the freezer and warm to room temperature, wrapped in foil before opening.
4. Prior to labeling the sections, make wells to contain small volumes of solutions on the sections by ringing them with a wax pen. Immerse the whole slide in a large volume of liquid for washing steps (e.g., in a coplin jar). For blocking, antibody incubations, and the detection steps, dry off the slides around the rings of wax, taking care not to dry out the sections. Place the slides on a raised platform in an airtight box and carefully pipet the solutions onto the sections. Place

strips of tissue paper on the bottom of the box, thoroughly wetted with buffer, and then put the lid on the box. This step is essential to create a humid atmosphere in the box to prevent evaporation of the small volumes of antibody solutions.

5. Labeling is carried out using the same solutions as the wholemount protocol for embryos described above. Rehydrate the sections in a large volume of PBS with several changes of buffer. Dry off the slides and apply the blocking solution (10% goat serum) and incubate at room temperature for 1 h. If the sections are small and placed close together, a volume of 10–20 µL is sufficient. After incubation, remove the block by inverting and draining the slides and replace with antibody solution diluted as described above. It may be necessary to increase the concentration of the primary antibody or use an enhancement technique as described in **Note 4** to amplify the signal. It is possible to carry out shorter incubations on sections, but once again, this is flexible, generally a minimum of 1 h at room temperature or overnight at 4°C. After the primary antibody incubation, wash thoroughly in PBT, changing the wash several times over 1 h at least. Increased washing times can reduce background staining. If the specimen is older than 30 h, it will be necessary to inactivate endogenous peroxidases as described in **Subheading 3.2.** After thorough washing, dry slides and apply the secondary antibody and incubate as for the primary antibody. The specimen must then be washed extensively before the detection step. Dry the slides and place in the humidified box and overlay with a small volume of DAB solution with H_2O_2 added (*see* **Subheading 2.**) and monitor the reaction using a low-power microscope. Take care to contain the DAB solution within the box and to inactivate any spilt DAB by washing the box with a weak bleach solution after use. To stop the reaction, pipet off the used DAB solution and inactivate in bleach, place the slides into a large volume of PBS, and change several times. The sections are cover-slipped using 70% glycerol or cleared in graded alcohols and mounted in Permount or DPX.

3.4. Double-Labeling Techniques

It is possible to label both wholemount specimens and sections with two different antibodies. If the antibodies are two different classes, e.g., one a rabbit polyclonal and the second a mouse monoclonal, it is possible to combine the two antibodies in the primary incubation. The secondary antibodies are applied sequentially, developing the first one with DAB alone for a brown product and the second with DAB plus nickel and cobalt to get a black reaction product (**Fig. 1C**). Double labeling works particularly well with fluorescent-conjugated secondary antibodies and these can be applied at the same time (**Fig. 1D**). Fluorescent preparations are best visualized using confocal microscopy, and the zebrafish is particularly amenable to this technique. If the two primary antibodies are of the same class, it is necessary to apply and develop the first antibody, wash extensively after the DAB step and treat with methanol–H_2O_2 as described in **Subheading 3.2.**, to inactivate the peroxidase attached to the secondary antibody. The second primary antibody is then applied and developed using DAB plus nickel and cobalt, resulting in a contrasting black reaction product.

3.5. Mounting and Sectioning of Wholemount Labeled Specimens

1. Whole embryos can be dissected in 70% glycerol to remove the yolk for flat mounting or removal of the skin and eyes to reveal internal labeling. This is achieved by pinning embryos to dishes coated with Sylgard silicone elastomer (BDH) with fine pins and using sharpened tungsten needles and fine forceps to dissect. Embryos are mounted in wells formed by attaching layers of one to four 22 × 22-mm cover slips on both ends of a long 22 × 64-mm cover slip with DPX or glue. A long cover slip is placed on top and the embryo can be viewed at higher power. The top cover slip can be held in place with nail varnish. Generally, embryos can be stored indefinitely in 70% glycerol in 24-to 48-well plates at 4°C.
2. It is often informative to section wholemount embryos after labeling. To maintain tissue morphology, embed embryos in JB4 resin (Agar Scientific) according to the manufacturer's instructions and cut 10-µm sections using a tungsten knife on a Jung 2055 Autocut microtome (Leica, Milton Keynes, UK) *(1)*. Collect plastic sections individually on a drop of water on an uncoated glass microscope slide and dry on a hot plate at 70°C. Sections may be counterstained with Toluidine Blue and cover-slipped directly with Permount or DPX. Embryos can be embedded immediately after labeling or if stored in 70% glycerol, rehydrated through graded glycerols (50%, 30%), and washed in PBS several times prior to embedding.

4. Notes

1. Generally, formaldehyde-based fixatives described in **Subheading 2.** are the best for immunohistochemistry and maintaining tissue morphology; however, some antigens/epitopes are sensitive to aldehydes and alternatives must be used, such as 100% acetone or methanol. Aldehyde fixation can also impair penetration by the primary antibody which can be avoided by using noncrosslinking fixatives such as trichloroacetic acid (TCA). To increase antibody penetration, zebrafish embryos older than 40 h require a permeabilization step. This can be achieved by enzymatic treatment (e.g., with trypsin or proteinase K) or by treatment with distilled water and solvents (*see* **Note 7**). The final choice of fixative and permeabilization will reflect the conditions required by the primary antibody and the age of the tissue.
2. When testing new antibodies, it is necessary to include a number of controls for specificity. Ideally, in one control, the primary antibody should be replaced with the preimmune serum or hybridoma supernatant. If preimmune serum is not available, then nonimmune sera from the same species can be used. Additional controls include using the secondary antibody alone and no antibodies to determine nonspecific background and the levels of endogenous enzymatic activity when

using secondary antibodies conjugated to HRP or AP. It is also useful to include a known antibody of the same class as a positive control because the test antibody to establish that the method is working.
3. Problems with nonspecific backgrounds can be alleviated by extending washing steps, adding carrier proteins such as 1% goat serum, 1% bovine serum albumin, or 1% milk powder to all washes, diluting the primary antibody further and using an enhancement technique to amplify the specific signal if necessary, changing the secondary antibody or preabsorbing the secondary antibody using zebrafish powder or whole fixed embryos.
4. Secondary antibodies are available commercially (e.g., from companies such as Sigma, Jackson Laboratories, and Vector Laboratories).
5. As DAB oxidizes in contact with air, there is little advantage to extending the reaction time beyond 30 min. If the signal is very weak, the experiment should be repeated using a higher concentration of primary antibody or an enhancement step, as described in **Note 7**. There are alternative substrates to DAB such as 3-amino-9-ethyl-carbazole (AEC) available; however, often these are less sensitive and not permanent. Most substrates are potential carcinogens and teratogens; they must be treated with care and inactivated after use. DAB is oxidized rapidly in the presence of a weak bleach solution. All pipet tips, tubes, plates, and surfaces that come into contact with DAB must be inactivated using a weak bleach solution over several hours and then disposed of appropriately.
6. The signal can be enhanced by taking advantage of the high binding affinity of the glycoprotein avidin for biotin. Biotin groups are attached to the secondary antibody. A third protein complex is formed by binding avidin molecules to enzymes such as HRP via further biotin molecules. This complex will then bind to biotin groups on the secondary antibody, amplifying the signal many times. The individual components are available commercially or can be purchased as part of a kit such as Vectastain (Vector Laboratories, Peterborough, UK) and have been used successfully both on zebrafish wholemounts and sections.
7. Permeabilisation: As an example of an alternative to trypsin treatment, the following method has been used for early embryos and can also be applied to larvae. After fixation with formaldehyde-based fix and washing in PBS as described in **Subheading 3.**, wash in dH$_2$O for 5 min, soak for 7 min in 100% acetone at –20°C, rinse in dH$_2$O once, then twice in PBT, and continue with the protocol from the blocking step *(11)*.
8. As an alternative to cryosectioning, a freeze substitution method to allow embedding in low-melting-point paraffin wax for sectioning which preserves tissue morphology and antigenicity, has been described elsewhere *(12)*.
9. The best coating for glass slides to attach zebrafish cryosections is TESPA (3-aminopropyltriethoxysilane; Sigma).

Acknowledgments

I would like to thank all my colleagues in the laboratories of Steve Wilson and Nigel Holder for their input into the development of these methods. In particular, I would like to acknowledge Lisa Leonard who originally developed the method for wholemount labeling of zebrafish larvae. Thanks to Sam Cooke, Dominic Delaney, Tom Schilling, and Steve Wilson for their comments on this manuscript. Research in Steve Wilson's laboratory is supported by the BBSRC and the Wellcome Trust.

References

1. Macdonald, R., Scholes, J., Strähle, U., Brennan, C., Holder, N., Brand, M., et al. (1997) The Pax protein Noi is required for commissural axon pathway formation in the rostral forebrain. *Development* **124,** 2397–2408.
2. Macdonald, R. and Wilson, S. W. (1997) Distribution of Pax6 protein during eye development suggests discrete roles in proliferative and differentiated visual cells. *Dev. Genes Evol.* **206,** 363–369.
3. Macdonald, R., Xu, Q., Barth, K. A., Mikkola, I., Holder, N., Fjose, A., et al. (1994) Regulatory gene expression boundaries demarcate sites of neuronal differentiation and reveal neuromeric organisation of the zebrafish forebrain. *Neuron* **13,** 1039–1053.
4. Wilson, S. W., Ross, L. S., Parrett, T., and Easter, S. S., Jr. (1990) The development of a simple scaffold of axon tracts in the brain of the embryonic zebrafish, *Brachydanio rerio*. *Development* **108,** 121–145.
5. Patel, N. H., Martin-Blanco, E., Coleman, K. G., Poole, S. J., Ellis, M. C., Kornberg, T. B. and Goodman, C. S. (1989) Expression of engrailed protein in arthropods, annelids and chordates. *Cell* **58,** 955–968.
6. Harlow, E. and Lane, D. P. (1988) *Antibodies: A Laboratory Manual.* Cold Spring Harbor Laboratories, Cold Spring Harbor, NY.
7. Ishikawa, Y., Zukeran, C., Kuratani, S., and Tanaka, S. (1986) A staining procedure for nerve fibers in whole mount preparations of the medaka and chick embryos. *Acta Histochem. Cytochem.* **16,** 775–783.
8. Ishikawa, Y. and Hyodo-Taguchi, Y. (1994) Cranial nerves and brain fiber systems of the medaka fry as observed by a whole-mount staining method. *Neurosci. Res.* 19, 379–386.
9. Westerfield, M. (1995) The Zebrafish Book, 3rd ed., University of Oregon Press, Eugene, OR.
10. Vischer, H. A. (1989) the development of lateral-line receptors in Eigenmannia (Teleosti, Gymnotiformes) I. The mechanoreceptive lateral-line system. *Brain Behav. Evol.* **33,** 205–222.
11. Metcalfe, W. K., Myers, P. Z., Trevarrow, B., Bass, M. B., and Kimmel, C. B. (1990) Primary neurons that express the L2/HNK–1 carbohydrate during early development in the zebrafish. *Development* **110,** 491–504.
12. Griffin, K. (1993) Gentle fixation by freeze substitution gives excellent histological results with zebrafish embryos. *Zebrafish Sci. Monitor* **2(3),** 2–3.

8

Immunohistochemistry of *Xenopus* Embryos

Carl Robinson and Matt Guille

1. Introduction

Immunohistochemistry is a very powerful technique for determining both the tissue-specific and subcellular location of endogenous and exogenous proteins within an embryo. The technique is relatively simple and when used on its own or in conjunction with other immunological techniques, such as Western blotting, immunoblocking, and supershift assays (*see* Chapter 15), can provide a large amount of information about the potential function or regulation of a given protein in a relatively small amount of time.

The basic technique utilizes an antibody that has been raised in a particular species (usually mouse or rabbit) against the protein of interest. This "primary antibody" is then used to identify the endogenous location of the protein. Embryos that have been fixed and had nonspecific binding sites preblocked are incubated (usually overnight) with the primary antibody, excess antibody is then washed off and the embryos are incubated with a second antibody (the secondary antibody), usually raised in a goat or sheep, that recognizes a particular class of immunoglobulin. The secondary antibody is conjugated to an enzyme such as alkaline phosphatase or horseradish peroxidase. The antigen–antibody complex can then be incubated with an analog of the enzyme's natural substrate, such as BM Purple (Boehringer Mannheim, Lewes, UK) or diaminobenzidine (DAB), to generate a colored precipitate (*see* **Fig. 1**).

There are two main types of antibodies that can be used. Monoclonal antibodies are produced in mouse hybridomas and recognize only one epitope of a particular protein. Polyclonal antibodies are raised against whole proteins or regions of proteins or peptides and usually recognize multiple epitopes of the protein. Both types of antibody have advantages, disadvantages and particular applications (*1–3*); however, the best results in wholemount experiments are

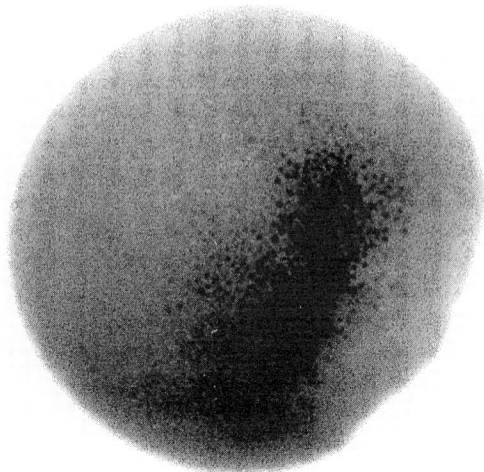

Fig. 1. Visualizing ectopic protein expression by immunodetection of an epitope tag. A *Xenopus* embryo was injected with 50 pg of mRNA encoding a Myc-tagged *Xenopus* nuclear protein. Once it had developed to stage 9, the embryo was fixed, bleached, and immunohistochemistry was performed using the 9E10 monoclonal antibody (Santa Cruz Biotechnology) and an anti-mouse Fab HRP-conjugated secondary antibody. The embryo was then developed using Sigma Fast DAB + metal enhancer. The dark staining within the embryo indicates where the tagged protein has been expressed within the embryo.

generally obtained with polyclonal sera raised against as large a proportion of the original protein as possible. A number of companies will produce both types of antibody to your protein of interest commercially, and in the current climate of animal welfare legislation, it is probably preferable and cost-effective to use this route of production.

It is essential that the antibody is characterized before use. The easiest way to characterize an antibody, especially if you do not know where the protein is expressed, is by Western blotting. This, in essence, is the same as the wholemount protocol, but the proteins are first denatured, separated by sodium dodecyl sulfate–polyacrylamide gel electrophoresis (SDS–PAGE) and transferred on to a nitrocellulose membrane before they are detected using antiserum. Probably the most important controls in immunohistochemistry are to show that the preimmune serum from the same animal as the immune serum does not contain antibodies that bind to proteins within the embryo, and that the antibody detects a single protein of the expected size in a Western blot.

Immunohistochemistry can also be used for checking the location of exogenous proteins. The gene encoding the protein of interest can be engineered so that an N- or C-terminal epitope tag is expressed upon translation of exogenous mRNA. Changes in the subcellular location of the expressed protein can then be followed by using an antiepitope antibody without any interference or masking from the endogenous protein (*see* **Fig. 1**). This can be especially useful if the protein has been mutated so that it mislocates, for example, if you have altered the protein's nuclear localization signal so that it should not enter the nucleus. The small size of the epitope tag (usually about eight amino acids) means that it is much less likely to interfere with the correct function or regulation of the protein when compared to relatively large protein tags like green fluorescent protein (GFP) or β-galactosidase.

In this chapter, we will outline a basic technique for Western blotting, as well as techniques involved in whole mount immunohistochemistry of *Xenopus* embryos. A detailed discussion of the production of types, choice and production of primary antibody will not be included and we refer readers to **refs. *1–3*.**

2. Materials

2.1. Fixing and Storing Xenopus Embryos

1. Sharp forceps.
2. MEMFA: 0.1 M MOPS, pH 7.4, 2 mM ethylene glycol-bis[β-aminoethyl-ether]-$N,N,N'N'$-tetraacetic acid (EGTA), 1 mM MgSO$_4$, 3.7% formaldehyde (prepared just prior to use).
3. Glass vials (15 mL).
4. Methanol.
5. Cut off 1-mL pipet tips.

2.2. Rehydration and Bleaching.

1. PBST: Phosphate-buffered saline (PBS) pH 7.4 (Sigma, Poole, UK) containing 0.1% Tween-20.
2. 75% Methanol, 50% methanol, and 25% methanol in PBST.
3. Bleaching solution: 5% formamide, 0.5X SSC, 10% H$_2$O$_2$. Make up from 50% formamide, 5X SSC diluted with water, and finally, H$_2$O$_2$ is added to a final concentration of 10%. *Do not* add the H$_2$O$_2$ directly to the formamide–salt sodium citrate (SSC) solution, as the mixture is explosive.

2.3. Xenopus Wholemount Immunohistochemistry

1. PBST as above.
2. Nonimmune serum, heat inactivated (Sigma).
3. Primary antibody (*see* **Note 1**).
4. Secondary antibody.
5. Alkaline phosphatase (AP) react buffer: 100 mM Tris–HCl, pH 9.5, 100 mM NaCl, 5 mM MgCl$_2$.

6. BM Purple (Boehringer Mannheim).
7. Trsi EDTA (TE): 10 mM Tris–HCl, pH 8.0, 1 mM EDTA.
8. Sigma Fast diaminobenzidine (DAB) + metal enhancer (Sigma).

2.4. Embedding and Sectioning Using a Vibratome

Although alternative methods for sectioning are available (in most standard texts), we have found that this particular one gives good signals when very small amounts of protein are present, because of the thickness of the sections obtained.

1. Gelatin–albumin mix: dissolve 2.2 g of gelatin in 450 mL of PBS by heating to 60°C. Cool the mixture, add 135 g of egg albumin, and dissolve (this usually takes about 1–2 h, but can take overnight). Add 90 g of sucrose, dissolve, and store aliquots at –20°C. Remove an aliquot from –20°C when required; this can be stored at 4°C for a few days.
2. Universal tubes (30 mL).
3. Glutaraldehyde (25%, Sigma).
4. Superglue.
5. Vibratome series 1000 sectioning machine.
6. Microscope slides and large cover slips.
7. Fine paint brush.
8. 90% glycerol.
9. Clear nail varnish.

2.5. Characterizing Antibody Specificity by Western Blotting

1. Sets of 25 embryos (*see* Chapter 10 for embryo preparation).
2. Embryo extraction buffer: 10 mM HEPES pH 8.5, 2 mM MgCl$_2$, 1 mM EDTA, 10 mM β-glycerophosphate, 1 mM dithiothreitol (DTT). Immediately before use, add one Complete mini EDTA-free protease inhibitor tablet to 5 mL of the buffer. This buffer can be used for about 1 wk if stored at 4°C.
3. 1,1,2-Trichlorotrifluoroethane (Freon).
4. Sodium dodecyl sulfate (SDS) loading buffer (2X): 100 mM Tris–HCl, pH 6.8, 4% SDS, 200 mM DTT, 0.1% bromophenol blue, 20% glycerol. Store 1 M DTT in water at –20°C and add just prior to use.
5. SDS–PAGE gel equipment and solutions (*see* **ref. 4**).
6. Western transfer buffer: 20 mM Tris base, 150 mM glycine, 0.1% SDS, 20% methanol.
7. Western transfer tank, including fiber pads and gel holder.
8. Whatman 3MM chromatography paper (Maidstone, UK).
9. Nitrocellulose membrane.
10. Tris-buffered saline Tween (TBST): 20 mM Tris–HCl, pH 7.5, 10 mM NaCl, 0.05% Tween-20.
11. MBS: TBST, 2–4% nonfat milk.
12. Primary antibody.

13. Secondary antibody.
14. Substrate solutions (*see* **Subheading 2.3.**).

3. Methods

3.1. Fixing and Storing Xenopus Embryos (modified from ref. 5)

1. Devitelline the embryos using very sharp forceps and pierce the blastocoel or archenteron with a microinjection needle to avoid accumulation of nonspecific background.
2. Transfer the embryos into MEMFA (10 mL) in small glass vials (15 mL) and incubate for 30 min at room temperature.
3. Replace the MEMFA by methanol (10 mL) twice, and leave the embryos in the second 10 mL at room temperature for 30 min.
4. Replace the methanol with a fresh aliquot and store at $-20°C$. The embryos are stable for at least 6 mo under these conditions

3.2. Rehydration and Bleaching.

1. Transfer 10–20 embryos to an Eppendorf tube (*see* **Note 2**) and remove the methanol very carefully. All subsequent washes are 1 mL unless stated.
2. Rehydrate the embryos by replacing the methanol sequentially with 75% methanol, 50% methanol, and 25% methanol in PBST and then finally PBST for 5 min each.
3. Suspend the embryos in bleaching solution and place the Eppendorfs on their sides on a light box for 5–10 min or until the pigment is no longer visible (*see* **Note 3**).
4. Remove the bleaching solution and wash the embryos three times in PBST for 5 min each wash.

3.3. Xenopus Wholemount Immunohistochemistry

1. Fix and bleach embryos as described above.
2. To block nonspecific binding sites for immunoglobulins, replace the final wash with 1 mL of PBST + 20% nonimmune serum (*see* **Note 4**) and incubate for 2 h at room temperature with gentle shaking.
3. Exchange the PBST/nonimmune serum with a fresh aliquot of PBST/nonimmune serum and add 1:1000 dilution of the primary antibody (*see* **Note 5**). Incubate overnight at room temperature with gentle shaking.
4. Wash off excess antibody with 5×1 mL aliquots of PBST, 1 h per wash, at room temperature with gentle shaking.
5. Reblock (*see* **Note 6**) the embryos by replacing the final wash with 1 mL of PBST/nonimmune serum and incubate for 2 h at room temperature with gentle shaking.
6. Exchange the PBST/nonimmune serum with a fresh aliquot of PBST/nonimmune serum and the secondary antibody (*see* **Note 7**). Incubate overnight at room temperature with gentle shaking.

7. Wash off excess antibody with 5 × 1 mL aliquots of PBST, 1 h per wash, at room temperature with gentle shaking.
8. Develope alkaline phosphatase secondary antibody conjugates. Replace the final PBST wash with 3X washes of AP react buffer for 5 min each wash, and then add 1 mL of BM purple (*see* **Notes 8** and **10**). Stop the reaction by transferring the embryos to 20 mL of TE.
9. Developing horseradish peroxidase linked secondary antibody conjugates. Make up 10 mL of Sigma Fast DAB + metal enhancer (*see* **Notes 9** and **10**) and transfer the embryos to about 2.5 mL of Fast DAB. Stop the reaction by transferring the embryos to 20 mL of TE.
10. Refix the embryos in MEMFA as described above and store at –20°C.

3.4. Double Staining

Double labeling of wholemount *Xenopus* embryos is possible. The easiest way to achieve this is to use primary antibodies that have been raised in different species or are of different classes and combine the two antibodies in the primary antibody incubation. The embryos are then stained with one antibody per substrate and then restained with a second combination. A very useful way to avoid masking one color reaction with a second is to use one of the secondaries that is conjugated to a fluorescent molecule as an antibody. This can then be visualized using the appropriate fluorescence attachment to the microscope.

3.5. Embedding and Thick Sectioning Wholemount Embryos (Based on ref. 6)

1. Rehydrate the embryos as described above (*see* **Note 11**).
2. Add 2 mL of gelatin–albumin mix to the embryos and leave overnight at 4°C (*see* **Note 12**).
3. Into the cap of a plastic universal tube add 2 mL of gelatin–albumin mix, and add a second 2 mL to the tube itself.
4. Add 200 µL of 25% gluteraldehyde to the gelatin-albumin mix in the cap and stir in with the pipet tip.
5. Place the embryos on top of the mixture.
6. Add 200 µL of 25% gluteraldehyde to the gelatin-albumin mix in the tube, stir well, and quickly pour over the embryos.
7. The mixture sets in about 10 min, and can be sectioned within 15 min; however, the embryos will section much better if the mixture is covered by cling film and left overnight or the weekend at 4°C.
8. Cut a square block, containing the embryos, from the embedding mixture, glue to a vibratome chuck, and allow to set.
9. Place the chuck in the vibratome, and add water to the reservoir until it is just above the level of the block.
10. Level the top of the block off, cut 30- to 60-µm sections and place on a microscope slide using a fine paint brush (place about six sections on each slide). Repeat the procedure until the whole of the embryo has been sectioned.

11. Remove excess water and place 150 µL of 90% glycerol to one side of the sections.
12. Gently place a cover slip over the embryos so that they become covered in the 90% glycerol.
13. Seal the cover slip to the slide using clear nail varnish. The sections can now be photographed.

3.6. Characterizing the Antibody Specificity by Western Blotting

Characterizing the antibody–antigen interaction by Western blotting consists of a three-step process: The embryo extracts (containing the antigen(s)) are separated by SDS–PAGE, transferred onto a nitrocellulose membrane, and then the protein of interest is detected by incubation with the primary antibody; then, after washing the secondary antibody is added and, finally, the enzyme on the secondary antibody is detected.

1. Homogenize 25 embryos (*see* **Note 13**) in 250 µL of extraction buffer. Remove the yolk proteins by adding an equal volume of 1,1,2-trichlorotrifluoroethane (Freon) *(7)*, vortex for 10 s and centrifuge at 21,000*g* for 3 min. Transfer the top aqueous layer to a new tube (*see* **Note 14**).
2. Add 200 µL of 2X SDS loading buffer, heat to 100°C for 3 min, and centrifuge for 5 min at 21,000*g*.
4. Separate the proteins by SDS–PAGE (*see* **Note 15**) as described in **ref. 4**). As a general rule, 20 µL of embryo extract in SDS loading buffer per lane, which is the equivalent of one whole embryo, is ideal for a mini gel system like the Atto or the Bio-Rad Mini PROTEAN II.
5. Remove the gel from the SDS–PAGE apparatus and soak in Western transfer buffer for 30 min at room temperature.
6. Cut two pieces of 3MM paper and a piece of nitrocellulose to the same size as the gel, and presoak them and two fiber pads in Western transfer buffer.
7. Place a fiber pad on one side of the gel holder and then place one piece of 3MM paper on top of this. Place the gel on top of the 3MM paper and lie the nitrocellulose membrane on it. Cover with the other piece of filter paper and remove any bubbles by rolling a test tube over the surface. Finally, add the second fiber pad and close the gel holder. Insert the gel holder into the transfer apparatus with the nitrocellulose filter facing the anode with respect to the gel. Fill the gel tank with Western transfer buffer and transfer the proteins onto the nitrocellulose membrane at 100 mA overnight or 400 mA for 2 h.
8. Block the membrane in 50 mL of MBS for 30 min at room temperature.
9. Replace the blocking solution with 10 mL of fresh blocking solution, add a 1:1000 dilution of the primary antibody (*see* **Note 5**), and incubate at room temperature for 2 h with gentle agitation.
10. Remove the primary antibody solution (*see* **Note 16**) and wash three times for 15 min with 100 mL of blocking solution.
11. Replace the final wash with 15 mL of fresh blocking solution, add the secondary antibody (*see* **Note 7**) and incubate for 1 h at room temperature with gentle shaking.

12. Remove the secondary antibody, wash three times for 15 min with 100 mL of blocking solution, and twice for 5 min with 100 mL of TBST.
13. Develop the blot as described in **Subheading 3.3.**, but using 2 mL of the appropriate substrate solution.

4. Notes

1. Store primary antibodies at −20°C in 200 µL aliquots. One aliquot should be thawed at a time and stored at 4°C. Freeze–thawing of antibodies reduces the titre and should be avoided.
2. Because of the delicacy of younger embryos (younger than stage 10.5), it is advisable that they are kept separate from the older ones for ALL subsequent manipulations.
3. Care must be taken when bleaching as the solution gradually heats up. If the embryos have not bleached within 10 min, the bleach should be removed and a fresh aliquot of bleaching solution should be added. Albino animals can be used to avoid the need for bleaching; however, they are more expensive and tend to be more difficult to keep compared to wild-type *Xenopus*.
4. To reduce background, use a nonimmune serum that corresponds to the animal that the secondary antibody was raised in, for example, if you have a secondary antibody raised in a goat, use goat serum.
5. Trial and error is the only way to work out how much antibody will be required for the particular antigen–antibody combination; however, 1:1000 dilution is a good starting point for polyclonal antisera. For monoclonal antibodies, use the manufacturer's recommended conditions. When testing a new serum, it is essential that a number of controls are carried out, these should include replacing the immune sera with the preimmune sera and verifying of the size of the detected proteins by Western blotting.
6. Reblocking the embryo after the primary antibody incubation is not essential, but it can significantly reduce the background staining upon development.
7. A number of companies (including Sigma, Vector laboratories, and Boehringer) produce a wide range of secondary antibodies that are conjugated to a number of enzymes, the most common of which are horseradish peroxidase (HRP) and alkaline phosphatase (AP). Initially, use the dilution of the antibody that the manufacturers suggest.
8. BM Purple, as its name suggests, gives a purple color; however, other reagents are available that give different colors. 5'-Bromo-4-chloro-3-indolyl-phosphate/Nitro blue tetrazolium (BCIP/NBT) and Fast Red (Boehringer Mannheim) give a blue color, 2-(4-Indophenyl)-5-(4-nitrophenyl)-3-phenyltetrazoline chloride (INT)/BCIP (Boehringer Mannheim) gives a red-brown color, and New Fuschin Red gives a red color.
9. A number of different substrates can be used to give different colors with horseradish peroxidase. Sigma Fast DAB + nickel enhancer gives a dark brown-black color, DAB gives a brown color, and 3-amino-9-ethylcarbazole gives a red color. DAB is a potent carcinogen and must be inactivated by the

addition of a 10% bleach solution after use. The developing time for both the HRP-linked and AP-linked antibodies can be as little as 30 s, and is best visualized at 3–4× magnification.
10. Nonspecific background staining can be avoided by increasing the wash times and volume of the washes, adding a small amount of nonspecific carrier to the wash solutions, decreasing the concentration the primary or secondary antibodies, and changing the secondary antibody. An increase in sensitivity can be achieved by increasing the amount of primary antibody and using a biotinylated secondary antibody followed by an avidin- or streptavidin-conjugated enzyme. Slightly overstaining wholemount embryos can be advantageous if they are to be sectioned, as the intensity of the staining decreases upon sectioning.
11. The embryos section better if they are stored in methanol for 2 or 3 d before being embedded.
12. Leaving embryos in gelatin–albumin mix overnight is not essential, but it does strengthen younger embryos when sectioning.
13. Collect sets of 25 embryos in a microfuge tube, remove as much of the water as possible without damaging the embryos, and store at –70°C.
14. To avoid contamination of the embryo extract with yolk proteins and insoluble cell debris that lies at the Freon–water interface, only take 200 µL of the embryo extract off the top of the freon.
15. Running identical samples on separate SDS–PAGE gels is important so that one gel can be transferred and Western blotted and one can be stained with Coomassie Blue to verify the overall protein composition.
16. The primary antibody solution can be stored at –20°C and can be reused about three or four times.

References

1. Catty, D., ed. (1988) *Antibodies: A Practical Approach*, IRL Press at Oxford University Press Oxford, Vol. I.
2. Catty, D., ed. (1988) *Antibodies: A Practical Approach*, IRL Press at Oxford University Press Oxford, Vol. II.
3. Harlow, E. and Lane, D. P. (1988) *Antibodies: A Laboratory Manual*, Cold Spring Harbor Laboratories, Cold Spring Harbor, NY.
4. Maniatis, T., Fritsch, E., and Sambrook, J. (1982) *Molecular Cloning: A Laboratory Manual*, Cold Spring Harbor Laboratories, Cold Spring Harbor, NY.
5. Harland, R. M. (1991) In situ hybridisation: an improved wholemount method for *Xenopus* embryos. *Methods Cell Biol.* **36,** 685–695.
6. Gove, C., Walmsey, M., Njjar, S., Bertwistle, D., Guille, M., Partington, G., et al. (1997) Over-expression of GATA-6 in *Xenopus* embryos blocks differential heart precursors. *EMBO J.* **16,** 355–368.
7. Wyllie, A. H., Laskey, R. A., Finch, J., and Gurdon, J. B. (1978) Selective DNA conservation and chromatin assembly after injection of SV40 DNA into *Xenopus* oocytes. *Dev. Biol.* **64,** 178–188.

9

Preparation and Testing of Synthetic mRNA for Microinjection

Wendy Moore and Matt Guille

1. Introduction

The misexpression and overexpression of proteins by injection of synthetic mRNA into *Xenopus* oocytes and, more particularly, embryos has provided the experimental basis for a considerable portion of our current knowledge of early vertebrate development (*1*, and references therein). The original experiments performed were designed to test the function of proteins in early development by expressing large quantities of the protein itself both in the cells and at the times when it would normally be expressed (overexpression) and in cells or at times when it would not (ectopic expression). These experiments was used to test the roles of both transcription factors and signaling molecules in *Xenopus* embryos.

More recently, the approaches taken using this technique have become more sophisticated; expression is now often targeted to particular regions of the embryo and the cells expressing the protein can be tracked. This can be achieved either by coinjection of a lineage tracer or by epitope tagging of the expressed protein followed by immunohistochemistry. The original type of experiment carried out by synthetic RNA injection was to test the effect of overexpression or ectopic expression as gain of function mutations (*2*). However, the lack of the ability to make specific null mutants in *Xenopus laevis* has recently been overcome, in part, by the overexpression of ingenious variants of the endogenous proteins under study. These dominant interfering mutant proteins disrupt the activity of their endogenous counterparts and effectively test gene inactivation. Examples include the expression of "dominant negative" versions of receptors (*3*), signaling molecules (*4,5*), enzymes (*6*), and transcription factors (*7*), all reviewed in **ref. 8**. In addition to expressing dominant negative versions of transcription factors in embryos, it has also proven possible to express both strongly activating and hormone-regulated versions of such factors. The design of dominant negative, activating, and regulated proteins and the vectors suitable for their expression as synthetic RNAs are discussed.

1.1. Interfering with the Action of Extracellular Signaling Molecules

The first dominant interfering mutant protein to be used in *Xenopus* was a truncated version of the fibroblast growth factor (FGF) receptor *(3)*. Like many receptors, its active form is a dimer *(9)*, and by expressing a large excess of a truncated receptor lacking the intracellular tyrosine kinase domain, it was possible to inactivate the endogenous receptor, titrating it out by forming dimers with the inactive dominant negative mutant. Clearly, when designing such dominant negative mutants, it is vital to have some assay to ensure that the form of the protein is genuinely a dominant negative. Amaya and co-workers were able to use an oocyte expression assay; they injected mRNAs encoding both the wild-type and mutant receptors and then tested for constructs which inhibited FGF-dependent Ca^{2+} release from the oocytes *(3)*. Using such an assay identified a control version of the truncated receptor vital in the final experiments and showed how much of an excess of the dominant negative form was needed to inactivate the endogenous receptor.

Such an approach has now been extended to a number of receptors *(10–18)*; however, it is necessary to treat the results obtained from this type of experiment with a degree of caution. Because many signaling molecules and their receptors are present in the embryo as families it has proven possible to inactivate the receptors for a range of family members with a single dominant negative receptor; for example a number of transforming growth factor-β (TGF-β) signaling pathways were inactivated by injection of the dominant negative activin type-II receptor into embryos *(19)*. Recently, other dominant interfering mutant approaches have been taken, in part, to overcome this problem. Joseph and Melton have made a mutant form of the Vg1 protein (a member of the TGF-β family) that acts in an apparently specific way to block Vg1 activity in the embryo *(5)*. They made a number of point mutants of the protein based on the known structure of other family members and then tested these, first for the lack of ability to induce mesoderm (mature Vg1 protein is a potent mesoderm inducer) and, then, by coinjection with the unmutated Vg1 mRNA, for the ability to block its activity in an animal-cap mesoderm induction assay (*see* Chapter 1). Similarly, dominant negative BMP-2, -4, and -7 ligands have been used to test the role of these signaling molecules during development *(4,13)*. It is clear that the processing of signaling molecules by enzymes is an important control step in development, and the overexpression of both a normal and a dominant negative version of a metalloprotease involved in the regulation of Spemann organizer activity has been used to test its function (6). In this case, the dominant negative version was able to be designed with reference to a related *Drosophila* protein. A mutation was introduced into the *Xenopus* protein which is a known antimorphic allele in the *Drosophila* homolog. Cotransfection of the unmutated and mutant forms of the protein into cultured cells was used to show that the mutant acts as a dominant negative form of the enzyme. This is by no means an exhaustive discussion of the ingenious ways in which proteins have been altered to interfere with signaling (e.g., there are constitutively active receptors also); however, these examples demonstrate the basic approaches.

1.2. Interfering with Transcription Factor Activity

In a similar way to that seen with the signaling molecules, a number of approaches have been taken to design altered versions of transcription factors. These, too are overexpressed, their activity tested during development and, thus, information gained regarding the activity of the endogenous factor. One common approach is to attempt to produce a dominant negative version of an activating transcription factor by fusing its DNA-binding domain with the transcription repression domain of the *Drosophila* Engrailed protein (20). This approach has been used to make dominant negative forms of a number of transcription factors, it has been used in cultured cells (21) and, subsequently, in *Xenopus* embryos. The first use in *Xenopus* was to test the role of the brachyury (Xbra) protein in development (7); Conlon and co-workers first mapped the transcription activation and DNA-binding domains of Xbra and then fused the DNA-binding domain to the Engrailed transcription repression domain. When synthetic mRNA was injected into embryos, they failed to complete gastrulation and mesoderm formation was severely inhibited. Injection of either the Xbra DNA-binding domain or the Engrailed transcription repression domain alone resulted in no altered phenotype and acted as controls. The other control consisted of rescuing the phenotype by coinjection of wild-type Xbra. Although experiments of this type have used linkage of the Engrailed repressor to DNA binding domains alone, Engrailed itself has a transcription activation domain (20) over which the repression domain dominates. It may, therefore, be possible to link the repression domain to an intact transcription factor and achieve repression, we are currently testing this premise and the initial experiments show it to be correct (C. Robinson and M. Guille, unpublished data). The Engrailed repressor domain (in common with several other transcription repressors) has been shown to act by binding the Groucho protein, which represses transcription directly (reviewed in **ref. 22**). It may be that more efficient dominant negative transcription factors can be obtained by fusion to the transcription repression domain of Groucho directly and cutting out the "middle man". In addition to the generally applicable method of making a dominant negative version of a transcription factor described above, other more specific methods have been used to similar effect *(23,24)*. A similar approach may be used to test the effect of transcriptional repressors during development, in this case the DNA-binding domain is fused to the strong transcription activator of VP16 *(25)*.

One disadvantage of all experiments involving RNA injection into embryos is that although expression can be directed to particular regions of an embryo it is not possible to control when it is expressed, as the injected RNA is translated immediately *(26)*. In an effort to overcome this problem for transcription factor expression, Kolm and Sive *(27)* and Tada and co-workers *(28)* have microinjected embryos with RNA, encoding a fusion protein between a transcription factor (MyoD and Xbra, respectively) and the hormone-binding domain of the glucocorticoid receptor. In the absence of a hormone, the transcription factor is unable to bind DNA and the embryos are unaffected. However, in the presence of dexamethasone, the transcription factor is released from the heat-shock apparatus, can bind DNA, and induces a phenotype. Thus, the nature of the activity of transcription factors and the time of their activation can be altered in *Xenopus* embryos. Experiments using these approaches are particularly suitable for testing the role of maternal transcription factors in development, as it is usually impossible to generate null mutants in these by any other method.

1.3. Following the Amount and Position of the Ectopically Expressed Protein

In many experiments (e.g., when the effect of expressing a number of different mutants of a factor is being compared), it is vital for the correct interpretation of results that the proteins produced by RNA injection are present in the equivalent cells in embryos and, made in similar amounts, of similar stability, and in the correct subcellular compartment. For these reasons, epitope-tagged versions of the proteins are often expressed in embryos. Although it is desirable to be able to follow the ectopically expressed protein some caution should be exercised as the tag may alter the activity of the protein; for example, the addition of five myc epitopes to the N-terminus of goosecoid turns it from a transcriptional repressor into an activator *(29)*. Tagging with myc epitopes is a popular method for tracking the expression of protein by immunohistochemistry, the commercially available antibody works well (*see* Chapter 8) and, in general, the activity of proteins seems to have been unaffected by the addition of the epitopes. Alternatives include the Flag tag marketed by Kodak and the hemagglutinin epitope *(28)*. The latter has been used in testing the stability of ectopically expressed proteins in embryos by Western blotting. Although these direct measurements of protein levels in the embryo are an ideal control, in situations when this is not possible (if tagging disrupts a protein complex, for example), then some attempt at checking the translational activity of the RNA to be injected should be made in vitro (below).

In summary, injection of synthetic mRNA into embryos has become an extremely flexible tool for testing the function of proteins in development. Although there are pitfalls in the use of this technique, the careful design of experiments and of controls has been used successfully by many groups to avoid them.

1.4. Vectors for the Preparation of Synthetic mRNA

There are a great number of vectors that have been made specifically for the preparation of stable, highly translated synthetic mRNA. The original one, pSP64T *(30)*, has many features that have been retained by those made recently. In essence the plasmid contains an SP6 RNA polymerase promoter upstream of the 5' untranslated region (UTR) of *Xenopus* β-globin (about 60 bp), a Bgl II site into that the open reading frame (ORF) of the protein which you wish to express is inserted, and then the 3' UTR of *Xenopus* β-globin (about 200 bp). This is followed by a synthetic poly A stretch of 30 nucleotides and a poly C stretch of similar length. After this, are a number of restriction enzyme sites for linearization of the plasmid prior to in vitro transcription. Transcription of the linearized plasmid in the presence of synthetic 5' Cap analog (below) then results in the production of a capped, polyadenylated RNA in which the ORF is surrounded by the UTRs from β-globin. This RNA is stable and highly translated in vitro, as well as in oocytes and embryos.

More recent plasmids are described on the World Wide Web and are being constantly updated (*see* http://vize222.zo.utexas.edu/Marker_pages/plasmids.html). The common modifications include multiple sites into which the ORF can be inserted, as well as the plasmids for fusion to the Engrailed repressor of transcription domain, the VP16 transcription activator and a hormone-binding domain. A series of plasmids is also available for the expression of high levels of protein by transfection into eucaryotic cells or by in vitro transcription.

2. Materials
2.1. Preparation of DNA for In Vitro Transcription

1. Plasmid DNA in RNase-free water (1 mg/mL) (*see* **Note 1**).
2. Restriction enzyme buffer.
3. Restriction enzyme to linearize the plasmid (*see* **Note 2**).
4. Water bath at the appropriate temperature for the above enzyme.

5. Buffer-saturated phenol–chloroform–isoamyl alcohol (50:50:1) (molecular biology grade, e.g., from Camlabs, Cambridge, UK).
6. Chloroform–isoamyl alcohol (50:1).
7. 3 M sodium acetate pH 5.0.
8. Ethanol at –20°C.
9. 70% Ethanol in RNase-free water.
10. RNase-free water.

2.2. Transcription Reaction

1. 5X Transcription buffer (Promega, Southampton, UK).
2. 2X NTP stock from Ambion or Pharmacia (*see* **Note 3**). SP6: 10 mM ATP, 10 mM CTP, 10 mM UTP, 2 mM GTP, 8 mM Cap analog. T7/T3: 15 mM ATP, 15 mM CTP, 15 mM UTP, 3 mM GTP, 12 mM Cap analog.
3. 100 mM dithiothreitol (DTT).
4. Ribonuclease inhibitor (e.g., RNasin, Promega).
5. SP6/T7/T3 RNA polymerase.
6. RNase-free DNase I.
7. Linearized template DNA (1 mg/mL).
8. 5 M Ammonium acetate.

2.3. Testing the Transcript

1. 1% agarose, 1X Tris-borate EDTA (TBE) gel (*see* **Note 4**).
2. In vitro translation kit (e.g., Promega nuclease-treated rabbit reticulocyte lysate).
3. 10% SDS-PAGE gel.
4. Phosphorimager and plates, if accurate quantitation of the translation products is required.

3. Methods

3.1. Preparation of DNA for In Vitro Transcription

1. Linearize 10 µg of plasmid DNA in a 50-µL reaction.
2. Add 50 µL of TE or water.
3. Extract once with 100 µL phenol–chloroform–isoamyl alcohol.
4. Extract once with 100 µL chloroform–isoamyl alcohol.
5. Add 10 µL 3 M sodium acetate.
6. Add 250 µL absolute ethanol (–20°C)
7. Vortex.
8. Spin at full speed in a microfuge for 20 min.
9. Carefully rinse the pellet in 70% ethanol.
10. Dry and resuspend in 10 µL RNase-free water.
11. Store at –20°C for up to 2 yr.

3.2. Transcription Reaction

1. Set the following reaction up at room temperature (*see* **Note 5**):

5X transcription buffer	5 µL
100 m*M* DTT	2.5 µL
RNasin	1 µL
2X NTP mix	12.5 µL
Linear template	2 µL
RNA polymerase	2 µL

2. Incubate at 37°C for 2 h.
3. Add 2 µL (20 units) of RNase-free DNase I.
4. Continue the incubation for a further 30 min.
5. Add 110 µL RNase-free water.
6. Add 15 µL 5 *M* ammonium acetate and mix.
7. Extract once with 150 µL phenol–chloroform–isoamyl alcohol.
8. Extract once with 150 µL chloroform–isoamyl alcohol.
9. Add 150 µL isopropanol.
10. Chill at –20°C for 30 min.
11. Spin at full speed in a microfuge for 20 min.
12. Carefully rinse the pellet in 70% ethanol.
13. Dry and resuspend in 50 µL RNase-free water.
14. Quantitate and test the RNA (below).
15. Store as small aliquots at –70°C for up to 2 yr.

3.3. Quantifying and Testing the RNA

1. Measure the optical density of a 100-fold dilution of the RNA at 260 nm and calculate the yield.
2. Confirm this by running an aliquot of the RNA (0.3–1 µg) on a TBE or TAE gel together with known standards (*see* **Note 6**).
3. Translate an aliquot of the RNA in an in vitro system (we use rabbit reticulocyte lysate). It is now unusual to prepare one's own in vitro translation mixes due to the costs involved; it is best to buy the commercial kit and follow the manufacturer's instructions. Take care when choosing the radioactively labeled amino acid to incorporate that that amino acid is reasonably represented in your protein.
4. Analyze the translation products by SDS-PAGE and autoradiography.
5. If you wish to have some comparison of activity between RNA samples for injection, then quantitative analysis of the translation products on a phosphorimager will provide this. It is, however, second best to checking the amount of protein synthesized in the embryos after injection.

4. Notes

1. The plasmid for in vitro transcription may be prepared by a variety of methods. However, if the method involves using RNase at any step, we use a final clean up of protease K digestion and an extra phenol–chloroform extraction step just to ensure that no contamination with RNase occurs.
2. Clearly, the enzyme chosen for linearization must not cut within the ORF of the protein you wish to express. For this reason some of the newer expression plas-

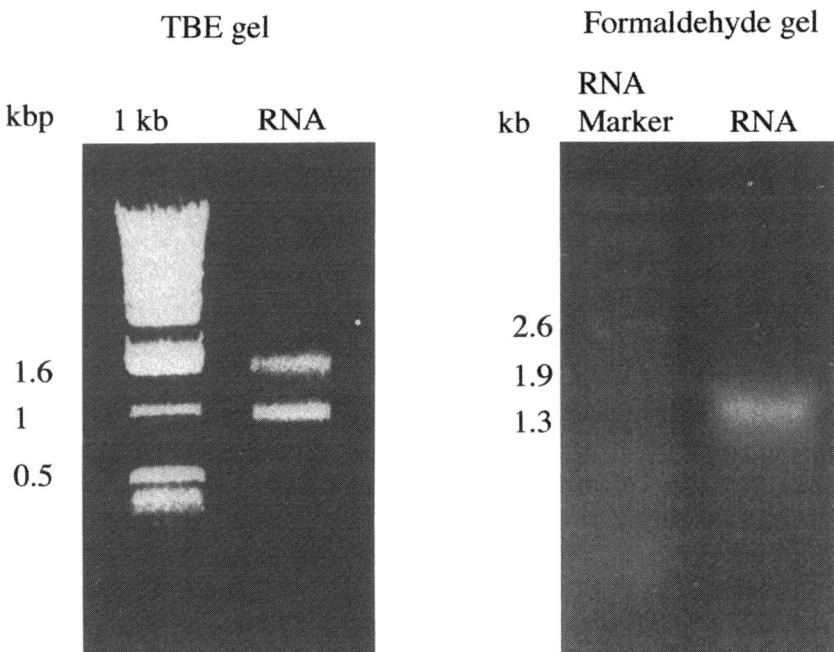

Fig. 1. Nondenaturing and denaturing agarose gel electrophoresis of a transcript produced in vitro. The ORF encoding the *Xenopus* homologue of nucleosome assembly protein 1 (xNAP-1) was inserted into pBUT2s. The resulting plasmid was linearised with Sfi I and transcribed with T3 RNA polymerase using a mMessage mMachine kit (Ambion, Austin, TX). One microgram of RNA was analyzed on either a 1% agarose, 1X TBE gel, or a 1% agarose denaturing gel containing formaldehyde *(31)*. The two bands seen under nondenaturing conditions become one species when denatured, strongly suggesting that they are a result of alternate secondary structures of the transcript.

mids have rarely cutting enzyme sites for linearization (e.g., Sfi I in pBUT2s and its derivatives). If using Sfi I, remember that the DNA must be very clean to cut to completion and that the enzyme works at 50°C.
3. Although the quantities of NTP for SP6 will work with T3/T7, the yield will be decreased and vice versa.
4. If the gel apparatus used to test the integrity of the RNA is ever used for other purposes, it is sensible to wash it thoroughly in 1 M NaOH, 0.1% SDS overnight before use in order to minimize RNase contamination. The gel must be made using components that are, as much as possible, free from such contamination.
5. The transcription buffer contains spermidine and may precipitate the DNA if the reaction is assembled cold or if the DNA is not added at the end. Commercial kits containing all the components necessary are available (e.g., Ambion m*M*essage

mMachine) but, although excellent, are quite expensive. We find that RNAs with similar activity are synthesized either with separate components or the kits.
6. The carry through of nucleotides from the transcription reaction occasionally gives inaccurate RNA concentrations when these are measured spectrophotometrically; thus, it is useful to confirm the concentration and integrity of the RNA on a gel. Do not be alarmed if multiple bands appear, these are usually the result of RNA secondary structure and may be checked by running a denaturing gel (*see* **Fig. 1**).

References

1. Heasman, J. (1997) Patterning the *Xenopus* Blastula. *Development* **124,** 4179–4191.
2. Hopwood, N. D. and Gurdon, J. B. (1990) Activation of muscle genes without myogenesis by ectopic expression of MyoD in frog embryo cells. *Nature* **347,** 197–200.
3. Amaya, E., Musci, T. J., and Kirschner, M. W. (1991) Expression of a dominant negative mutant of the FGF receptor disrupts mesoderm formation in *Xenopus* embryos. *Cell* **66,** 257–270.
4. Suzuki, A., Kaneko, E., Ueno, N., and Hemmati-Brivanlou, A. (1997) Regulation of epidermal induction by BMP2 and BMP7 signaling. *Dev. Biol.* **189,** 112–122.
5. Joseph, E. M. and Melton, D. A. (1998) Mutant Vg1 ligands disrupt endoderm and mesoderm formation in *Xenopus* embryos. *Development* **125,** 2677–2685.
6. Piccolo, S., Agius, E., Lu, B., Goodman, S., Dale, L., and DeRobertis, E. (1997) Cleavage of chordin by Xolloid metalloprotease suggests a role for proteolytic processing in the regulation of Spemann organizer activity. *Cell* **91,** 407–416.
7. Conlon, F. L., Sedgwick, S. G., Weston, K. M., and Smith, J. C. (1996) Inhibition of Xbra transcription activation causes defects in mespdermal patterning and reveals autoregulation of Xbra in dorsal mesoderm. *Development* **122,** 2427–2435.
8. Lagna, G. and Hemmati-Brivanlou, A. (1998) Use of dominant negative constructs to modulate gene expression. *Curr. Topics Dev. Biol.* **36,** 75–98.
9. Ullrich, A. and Schlessinger, J. (1990) Signal transduction by receptors with tyrosine kinase activity. *Cell* **61,** 203–212.
10. Smith, D. P., Mason, C. S., Jones, E. and Old, R. (1994) Expression of a dominant-negative retinoic acid receptor-gamma in *Xenopus* embryos leads to partial resistance to retinoic acid. *Rouxs Arch. Dev. Biol.* **203,** 254–265.
11. Suzuki, A., Thies, R. S., Yamaji, N., Song, J. J., Wozney, J. M., Murakami, K., et al. (1994) A truncated bone morphogenetic protein-receptor affects dorsal-ventral patterning in the early *Xenopus* embryo. *Proc. Natl. Acad. Sci. USA* **91,** 10,255–10,259.
12. Ataliotis, P., Symes, K., Chou, M. M., Ho, L., and Mercola, M. (1995) PDGF Signaling is required for gastrulation of *Xenopus*-Laevis. *Development* **121,** 3099–3110.
13. Hawley, S. H. B., Wunnenbergstapleton, K., Hashimoto, C., Laurent, M. N., Watabe, T., Blumberg, B. W., et al. (1995) Disruption of BMP signals in embryonic *Xenopus* ectoderm leads to direct neural induction. *Genes Dev.* **9,** 2923–2935.

14. Xu, Q. L., Alldus, G., Holder, N., and Wilkinson, D. G. (1995) Expression of truncated Sek–1 receptor tyrosine kinase disrupts the segmental restriction of gene-expression in the *Xenopus* and zebrafish hindbrain. *Development* **121,** 4005–4016.
15. Ulisse, S., Esslemont, G., Baker, B. S., Chatterjee, V. K. K., and Tata, J. R. (1996) Dominant-negative mutant thyroid-hormone receptors prevent transcription from the *Xenopus* thyroid-hormone receptor-beta gene promoter in response to thyroid-hormone in *Xenopus* tadpoles in-vivo. *Proc. Natl. Acad. Sci. USA* **93,** 1205–1209.
16. Aoki, S., Takahashi, K., Matsumoto, K., and Nakamura, T. (1997) Activation of Met tyrosine kinase by hepatocyte growth factor is essential for internal organogenesis in *Xenopus* embryos. *Biochem. Biophys. Res. Commun.* **234,** 8–14.
17. Glinka, A., Wu, W., Onichtchouk, D., Blumenstock, C., and Niehrs, C. (1997) Head induction by simultaneous repression of Bmp and Wnt signalling in *Xenopus*. *Nature* **389,** 517–519.
18. Frisch, A. and Wright, C. V. E. (1998) XBMPRII, a novel *Xenopus* type II receptor mediating BMP signaling in embryonic tissues. *Development* **125,** 431–442.
19. Hemmati-Brivanlou, A. and Melton, D. A. (1992) A truncated activin receptor inhibits mesoderm induction and formation of axial structures in *Xenopus* embryos. *Nature* **359,** 609–614.
20. Han, K. and Manley, J. (1993) Functional domains of the *Drosophila* engrailed protein. *EMBO J* **12,** 2723–2733.
21. Badiani, P., Corbella, P., Kioussis, D., Marvel, J., and Weston, K. (1994) Dominant interfering alleles define a role for c-myb in T-cell development. *Genes Dev.* **8,** 770–782.
22. Fisher, A. L. and Caudy, M. (1998) Groucho proteins: transcriptional corepressors for specific subsets of DNA-binding transcription factors in vertebrates and invertebrates. *Genes Dev.* **12,** 1931–1940.
23. Kim, J., Lin, J. J., Xu, R. H., and Kung, H. F. (1998) Mesoderm induction by heterodimeric AP–1 (c-Jun and c-Fos) and its involvement in mesoderm formation through the embryonic fibroblast growth factor Xbra autocatalytic loop during the early development of *Xenopus* embryos. *J. Biol. Chem.* **273,** 1542–1550.
24. Beck, C. W., Sutherland, D. J., and Woodland, H. R. (1998) Involvement of NF-kappa B associated proteins in FGF-mediated mesoderm induction. *Int. J. Dev. Biol.* **42,** 67–77.
25. Kessler, D. S. (1997) Siamois is required for formation of Spemann's organizer. *Proc. Natl. Acad. Sci. USA* **94,** 13,017–13,022.
26. Snape, A. M. and Smith, J. C. (1996) Regulation of embryonic-cell division by a *Xenopus* gastrula-specific protein kinase. *EMBO J.* **15,** 4556–4565.
27. Kolm, P. and Sive, H. (1995) Efficient hormone-inducible protein function in *Xenopus* laevis. *Dev. Biol.* **171,** 267–272.
28. Tada, M., OReilly, M. A. J., and Smith, J. C. (1997) Analysis of competence and of Brachyury autoinduction by use of hormone-inducible Xbra. *Development* **124,** 2225–2234.

29. Ferreiro, B., Artinger, M., Cho, K. W. Y., and Niehrs, C. (1998) Antimorphic goosecoids. *Development* **125,** 1347–1359.
30. Krieg, P. A. and Melton, D. A. (1984) Functional messenger RNAs are produced by SP6 in vitro transcription of cloned cDNAs. *Nucleic Acids Res.* **12,** 7057–7070.
31. Maniatis, T., Fritsch, E. F. and Sambrook, J. (1982) *Molecular Cloning; A Laboratory Manual*, Cold Spring Harbor Laboratories, Cold Spring Harbor, NY.

10

Microinjection into *Xenopus* Oocytes and Embryos

Matt Guille

1. Introduction

The ease of obtaining large numbers of *Xenopus laevis* eggs and oocytes together with their size and robust nature have made them a popular choice when microinjection of macromolecules into a cell is required for protein expression and modification studies, protein function analysis, RNA processing and stability analysis, gene expression investigations, and for studying development. Historically, such microinjection experiments were reported as early as 1971; thus, the advantages and pitfalls of using this technique are well known and have been the subject of review *(1,2)*. Here, the uses to which these techniques have been put will be outlined and then the preparation of oocytes and embryos, the microinjection apparatus, and the process of microinjection will be described in detail.

1.1. Oocyte Microinjection

The uses of microinjected *Xenopus* oocytes fall broadly into two categories: first, studies which use their ability to correctly modify and transport the products of exogenously expressed genes with a view to investigating the properties of these gene products themselves *(2,3)*. The second category involves studies on oocyte and early embryo metabolism or development. The first category includes microinjection of RNA encoding secreted proteins, transmembrane receptors and transporters for both expression cloning *(4–6)* and functional studies *(7–9)* and investigating sequences and mechanisms responsible for the correct intracellular localization *(10,11)* of proteins. The uses of injected DNA in this first category include testing the effect of DNA modification and structure *(12,13)* and chromatin structure *(14,15)* upon gene expression. In the latter type of "endogenous" study, microinjection of RNA, DNA,

From: *Methods in Molecular Biology, Vol. 127: Molecular Methods in Developmental Biology: Xenopus and Zebrafish* Edited by: M. Guille © Humana Press Inc., Totowa, NJ

antibodies *(16)*, and antisense oligonucleotides *(17,18)* has been used to study cell-cycle control and oocyte maturation *(17)*, the regulation of translation *(19)* and transcription *(20–22)*, and RNA metabolism *(23)* as well as to test the effect of depleting maternal gene products on the subsequent development of resulting embryos *(18)*.

The technique of oocyte injection has remained essentially unchanged for many years although advances have made it possible to achieve more consistent injection needles and volumes of injected material, automated injectors have even been produced *(24)*. In essence, ovaries are removed from a female and then the oocytes are either stripped manually from these or they are digested away from the surrounding membranes using collagenase. Once freed, the oocytes are examined and those suitable for injection are selected; generally, these will be healthy and stage VI. It is then usual to incubate these overnight prior to injection, as any that have been damaged during the removal process may take several hours to show signs that they are dying. If DNA injection into the nucleus is required, centrifugation of the oocytes may be carried out the next day to reveal the nuclei. Cytoplasmic or nuclear injections are then carried out under a stereomicroscope using a thin glass needle held in a micromanipulator. The injection volume is governed by an adjustable apparatus either of the "direct injection" type (*see* Chapter 12) or a gas valve (*see* below). Subsequent manipulations will depend on the exact nature of the experiment being undertaken.

1.2. Embryo Microinjection

Microinjection of RNA and DNA into fertilized eggs and cleavage-stage embryos of *Xenopus laevis* is, almost exclusively, used for developmental studies. Injection of synthetic mRNAs encoding proteins with roles central to early development has been a key technique in uncovering the molecular processes underlying patterning of the embryo *(25)*. The injection of DNA has also been used to test the effect of ectopic protein expression on development and, additionally, for analyzing cis-acting sequences in the regulatory regions of genes (both reviewed in Chapter 12).

The original type of experiment carried out by synthetic RNA injection was to test the effect of overexpression or ectopic expression as gain of function mutations *(26)*. However, the lack of the ability to make specific null mutants in *Xenopus laevis* has recently been overcome, in part, by the overexpression of ingenious variants of the endogenous proteins under study. These dominant interfering mutant proteins disrupt the activity of their endogenous counterparts and effectively test gene inactivation. Examples include the expression of "dominant negative" versions of receptors *(27)*, signaling molecules *(28,29)*, enzymes *(30)*, and transcription factors *(31)*. The design of dominant negative proteins and the vectors suitable for their expression as synthetic RNAs are discussed in Chapter 9. The relative merits of DNA and RNA injection for overexpression are discussed in Chapter 12, but, generally, RNA injection results in less mosaic expression than DNA but that expression is shorter-lived.

A variation on the usual overexpression experiments that has been applied to the study of secreted proteins with potential roles in regulating development is to inject synthetic RNA, encoding the protein of interest together with a lineage label into one embryo, and then to take an explant from that embryo and fuse it with an explant taken from an uninjected one. Analysis of injection-dependent changes in phenotype or gene expression in the recipient tissue often gives important information concerning the role of the secreted protein.

Briefly, microinjection experiments begin with the preparation of fertilized eggs, the type of experiment being undertaken will dictate the time of day chosen for embryo production. Natural matings are usually unsuitable for the sort of experiments considered here; therefore, it is necessary to fertilize the eggs with crushed testes. After removing the jelly coat from the embryos, their incubation temperature is usually lowered in order to slow development, thus lengthening the amount of time available for injection. The embryos are also transferred into buffer containing ficoll in order to stop cytoplasmic leakage from the injection wound. Injection may be into the fertilized egg, but more often will either be into one blastomere of the 2-cell embryo (thus allowing one-half of the embryo to act as control) or targeted to a particular region of the embryo at the 16- or 32-cell stage (*see* Chapter 12). The injection apparatus may be as for oocytes; however, if injection into specific cells of older embryos are needed, the very accurate "gas valve" type of injector is probably best. When targeted injections are carried out, it is essential that a lineage tracer is used to ensure that the injected material is expressed where one expects! There is a choice of lineage tracer (Chapter 12); however for most experiments, when the embryos will be analyzed by wholemount *in situ* hybridization or immunohistochemistry, mRNA- or DNA-encoding β-galactosidase is very effective (*see* Chapter 12).

In summary, oocyte and embryo injection are relatively simple techniques with a variety of applications. Together, they have been used to carry out a variety of key investigations not restricted to the field of development.

2. Materials
2.1. General for Microinjection

For a general view of a typical microinjection workstation, *see* **Fig. 1**.

1. Stereomicroscope equipped with a reticule micrometer and capable of at least 10× to 25× magnification (e.g., Nikon SMZ-U).

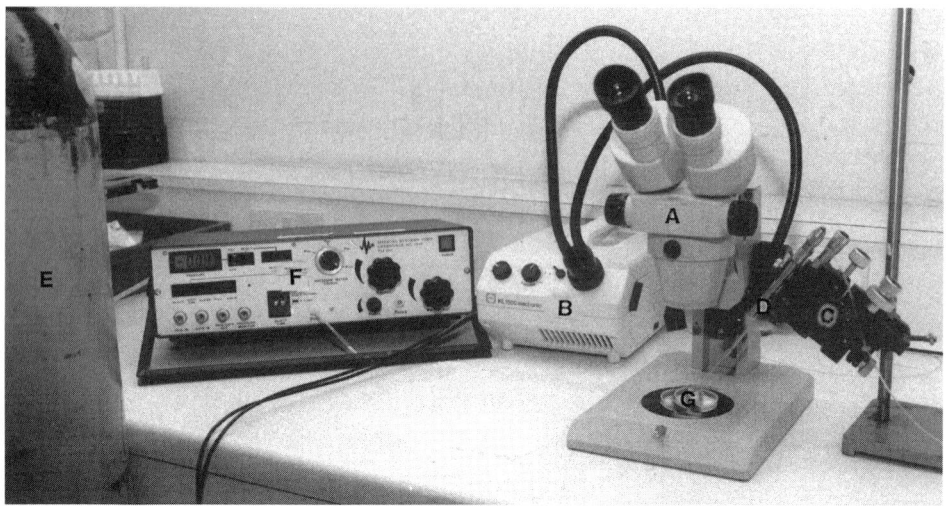

Fig. 1. A typical microinjection workstation for *Xenopus* work. **(A)** stereomicroscope; **(B)** cold light source; **(C)** micromanipulator; **(D)** needle holder; **(E)** nitrogen cylinder; **(F)** picoinjector apparatus; **(G)** grid for oocytes/embryos.

2. Cold-light source (e.g., Schott KL1500) equipped with a double "swan neck" or circular fiber-optic downlighter.
3. Micromanipulator (e.g., Sutter MM-33) mounted on a suitably stable base.
4. Micropipet puller (e.g., Sutter p-97).
5. Micropipets (1 mm outside diameter; 0.58 mm internal diameter without filament; e.g., Clark Electromedical GC100-10).
6. Watchmakers forceps (Dumont No. 5).
7. Pipet for moving oocyte/embryos. Either a Pasteur pipet with the end removed and blunted by fire polishing, or a Finn-type pipet with a cut off tip.
8. Microinjector (e.g., Medical Systems Corp. PLI100).
9. Incubation buffers for oocytes and embryos; e.g., Marc's modified Ringer's (MMR), Steinberg's solution (*see* Chapter 12) or modified Barths' saline (MBS): 88 mM NaCl, 1 mM KCl, 2.4 mM NaHCO$_3$, 15 mM HEPES–NaOH (pH 7.6), 0.3 mM CaNO$_3$, 0.41 mM CaCl$_2$, 0.82 mM MgSO$_4$, 10 U/mL sodium penicillin, 10 µg/mL streptomycin sulfate. To prepare MBS, make two stock solutions; all reagents should be of the highest available grade:

    High salt stock (solution A): HEPES (89 g), NaCl (128 g), KCl (2 g), NaHCO$_3$ (5 g). Dissolve these in 800 mL of distilled water, pH to 7.6 using NaOH, make to 1 L, filter sterilize and store at room temperature.

    Divalent cation stock (solution B): CaNO$_3$·4H$_2$O (1.9 g), MgSO$_4$·7H$_2$O (5 g), CaCl$_2$·6H$_2$O (2.25 g). Dissolve these in 1 L of distilled water, filter sterilize and store at room temperature.

Penicillin–streptomycin stock (solution C): Stabilized solution of 10,000 U/mL penicillin and 10 mg/mL streptomycin. Filter sterilize and store at –20°C. Available commercially (e.g., Sigma P4333).

To make 1X MBS, add 919 mL distilled water to 40 mL solution A and mix thoroughly; add 40 mL solution B and 1 mL solution C and mix (*see* **Note 1**). We find that this solution is stable for up to 1 mo when stored at room temperature or 14°C. 0.1X MBS is made identically, but using 4 mL of stocks A and B.

10. 3% (w/v) Ficoll in 1X MBS.
11. 60-mm Petri dishes into which 8-mm nylon mesh has been stuck using chloroform.
12. Temperature-controlled incubator (capable of at least 13°C to 25°C).

2.2. Obtaining Oocytes

1. Adult female *Xenopus laevis* (*see* **Note 2**).
2. Ethyl-*m*-aminobenzoate: 0.15% (W/V) solution in tap water (*see* **Note 3**).
3. Sterile instruments: scalpel, scissors and forceps.
4. Sutures (e.g., Ethicon mersilk 6/0).

2.3. Obtaining Embryos

1. Adult female *Xenopus laevis* (*see* **Note 2**).
2. Human chorionic gonadotrophin (HCG) (e.g., chorulon, from Intervet, Cambridge, UK).
3. 1 mL Syringe and 23 g, 3 cm needle.
4. Adult male *Xenopus laevis*, usually wild-caught.
5. 60% (v/v) Leibovitz L-15 medium, 10 units/mL sodium penicillin, 10 µg/mL streptomycin sulfate.
6. 2% Cysteine–NaOH, pH 7.8–8 in 1X MBS.

3. Methods.
3.1. Obtaining Oocytes

Oocytes may be obtained by one of two methods, either one or two lobes of the ovary may be removed under anesthetic and the frog allowed to recover, or the frog may be killed. In both cases, the procedure is likely to be covered by animal welfare legislation. As most experiments need fewer than 1000 oocytes, it is usually considered wasteful to kill a frog (which may well contain 30,000) to obtain these. Indeed, as oocytes from different females may vary considerably in their macromolecular content, it is sometimes desirable to be able to repeat experiments using oocytes from the same frog, clearly, the frog cannot be killed if this is the case. There are disadvantages to the reuse of frogs: it may be limited by law and, if this type of surgery with recovery is allowed, the training and facilities for it may be considered less costly than killing frogs. Both methods are described.

3.1.1. Removal of Total Ovary

1. Terminally anesthetize a female using 0.5% (w/v) ethyl-*m*-aminobenzoate.
2. Remove the head and use a dissecting needle to destroy the nerve tissue running down the spine.
3. Make an incision low in the ventral side of the abdomen to avoid major blood vessels, the ovary is visible immediately.
4. Quickly remove the lobes, cut free and rinse in 1X MBS.
5. Using forceps, pull the oocytes into small clumps (30–50 large oocytes) in Petri dishes containing MBS (*see* **Note 4**).
6. Replace the MBS with fresh and place in an incubator at 19°C.
7. As soon as possible, manually strip or collagenase treat the oocytes so that they can be cultured as individuals. Manual stripping is performed by holding a clump of oocytes with one pair of watchmaker's forceps and then running another pair over the surface of the clump in a gentle pinching movement. Sufficient force must be used to pull the large oocytes off without damage. This method is the first choice unless several hundred oocytes are needed. To collagenase treat, place the clumps of oocytes into 2 mg/mL collagenase and incubate with gentle agitation at 18–21°C for 2–4 h.
8. Select oocytes which are suitable for injection. Healthy oocytes have very well-defined pigmentation; that is, the border between the pigmented and nonpigmented halves of the oocyte is very clear. Similarly the pigment is evenly distributed in the animal half and the surface of the oocyte is smooth (*see* **Fig. 2**).
9. Culture these overnight in 1X MBS at 18–20°C, re-examine in the morning, and remove any unhealthy oocytes prior to injection.

3.1.2. Removal of Ovary Lobes

1. Anesthetize a female frog by immersion in 0.15% ethyl-*m*-aminobenzoate for 30 min.
2. Prepare a bed of water-saturated tissue and lie the frog onto this on its back. Ensure that the skin of the frog is kept damp at all times.
3. Carefully make a small incision in the lower abdominal wall through both the inner and outer skin.
4. Use blunt forceps to tease out one or two lobes of the ovary.
5. Tie a ligature around the base of the lobes and then cut them off, immediately tease them apart and then treat as in **step 5** of **Subheading 3.1.1.**
6. Flood the wound site with MBS.
7. Suture the inner body wall.
8. Suture the outer body wall.
9. Allow the frog to recover, ensuring that it is kept wet but taking care that no water is over the nostrils.
10. The frog should recover in 30–120 min, at which point return it to a tank on its own for subsequent observation for 48 h.
11. Remove the stitches when the wound is well healed (normally 7–14 d).

Microinjection into Xenopus Oocytes and Embryos

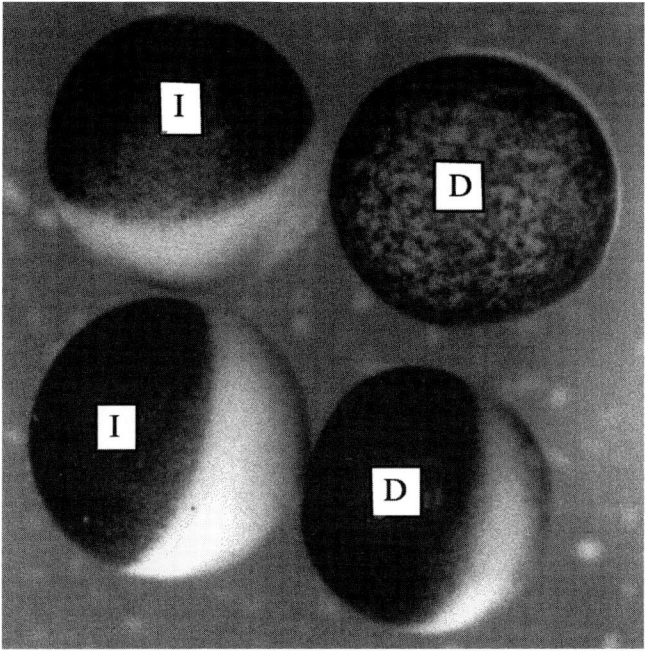

Fig. 2. Oocytes that are suitable and unsuitable for injection. Oocytes marked I are suited for injection; they are large and healthy with animal and vegetal poles well defined by pigment. Those marked D (discard) are ones that are unhealthy; note the uneven nature of the pigment or misshaped appearance.

3.2. Preparation of Embryos

1. On the evening before the microinjection, take two adult female frogs and inject them with 600–800 units of HCG (depending on their size). It may be easier to cool the frogs in iced water to slow them down prior to the injection. Inject into the dorsal lymph sac: On the dorsal side of the frog just above the junction of the back leg and the body is a semicircle of small slits. Below this, a membrane joins the outer skin to the body wall. Hold the syringe such that the needle is almost parallel to the skin and push it through just on the "leg side" of the marks, ensuring that it stays between the two layers. Penetrate the membrane under the marks and inject, ensure that the frog recovers safely if it has been cooled.
2. Keep the frogs overnight at 19–21°C.
3. Kill an adult male frog by tricaine anesthesia and remove the testes; these are found to the posterior of the liver, attached to the bright yellow fat bodies. Trim any extraneous tissue from them, rinse off any blood, and place in medium, either 1X MBS if they are to be used in the next 3 d or 60% (v/v) Leibovitz L-15 medium if they are to be kept for longer (up to 2 wk). Store at 4°C.

4. The next morning, the female frogs should be laying, if not try warming them to 23°C and placing them in bright sunshine.
5. Crush one quarter of a testis in 0.4 mL 0.1X MBS using forceps and then take the resulting suspension up and down in a 1-mL pipet tip until the tissue is well suspended.
6. Squeeze the female frogs gently but firmly around the abdomen so that they lay into a 9-cm Petri dish; move the frog so that the eggs fall into small groups (100–200 eggs). If the frogs squeak, you are squeezing too hard!
7. Using a pipet, squirt the crushed testes over each group of eggs and then shake the dish so that the eggs form a single layer. This procedure produces fertilization rates of >99% with most frogs.
8. Leave the eggs for 5 min, then rinse and incubate in 0.1X MBS.
9. After the eggs have turned so that the animal pole is uppermost (about 20 min postfertilization), dejelly them by pouring off the MBS and replacing it with the cysteine solution. Move the Petri dish vigorously and the eggs will detach themselves from the plate, pour them into a 50-mL conical bottomed tube, fill with cysteine, and shake gently. Occasionally let the eggs fall to the bottom of the tube and check whether they appear in direct contact with one another, any gap is the result of the remaining jelly. Once all are in contact when settled, wash them with at least four changes of 1X MBS and return them to a fresh Petri dish.
10. Transfer them to an incubator at 14°C to slow cell divisions and hence extend the period available for injection.

3.3. Microinjection

Microinjection is described generally for oocytes, with the specific procedures for oocyte nuclear and embryo injections given in **Notes 5** and **6**, respectively. Careful preparation for the experiment makes it much less fraught; for example, ensuring that there is sufficient gas for the injector and that injection needles (*see* **Note 7**) are available are best carried out on the day before. The timing of the experiment should also be considered. If embryos are to be manipulated at gastrula stages, it is often best to produce embryos and inject them in the evening, incubate them at 13–14°C overnight, and carry out the manipulations when fresh the next morning. This avoids having to carry out delicate manipulations when tired, the situation if a morning fertilization and injection are used.

1. Place a needle into the holder on the micromanipulator. If using the PLI100, do not overtighten the knurled nut that holds the needle in place, this occasionally cuts off the gas and no filling or injection can take place. For use of the Drummond injector, *see* Chapter 12.
2. Use watchmakers forceps to cut the end of the needle off, do this under the microscope at the point where the needle tip becomes less flexible. With practice this produces very consistent microinjection needles.

3. Set the injector to the required volume. For oocytes and embryos at the one- or two- cell stage, we use approximately 4 nL. Take up RNase-free water into the needle using the "fill" function and then, ensuring that the microscope zoom is set to 1× and that the end of the needle is in focus, inject a droplet into midair. A diameter of two divisions on our reticule micrometer is 4.1 nL, assuming the droplet to be spherical. In general, we find that with an injection pressure of 10 psi, injection times of 0.3 to 0.6 seconds are required, these conditions give excellent survival rates.
4. Fill the needle with the DNA or RNA to be injected (for DNA preparation, *see* **Note 8**, for RNA, *see* Chapter 9). Spin the solution for 5 min in a microcentrifuge to remove any debris that might block the needle and then pipet 1–2 μL onto a square of parafilm. With the needle tip well into the droplet, use the "fill" function. I usually keep watching the droplet from which the needle is being filled through the microscope in order to stop filling well before there is any chance of running out of the injection solution.
5. Prepare the oocytes for injection, transfer them to the grid in a Petri dish in 1X MBS, 3% ficoll. We usually find it best if groups of 100–200 oocytes are injected as a single batch when a 6-cm Petri dish is being used. This means that you have a good balance between stopping too often to refill the dish and not having the oocytes "pile up" over one another, causing injection problems. Remove the excess buffer from the dish until the top of the oocytes is just below the surface of the liquid. This stops them from moving around under pressure from the needle. It is important that they are not actually on the surface because, first, surface tension may burst them if the injection wound is sufficiently near to the top, and, second, the oocytes may dry out enough to damage viability.
6. Align the Petri dish grid under the microscope so that it is parallel to the apparent horizontal axis, this makes it easier to ensure that each oocyte has been injected.
7. Move the needle down toward the oocytes, I generally find that 15× magnification is good for injecting. Just before the needle enters the liquid, inject once to ensure that the resulting droplet is the correct size and that the needle has not become blocked.
8. Inject each oocyte, it is quite difficult to drive the needle tip through the surface of the oocyte when compared with an embryo (*see* **Note 9**). The oocyte will form a dimple under pressure from the needle, this should remain small until the point when it disappears as the needle penetrates (if not, the needle tip is too large), this usually occurs when the needle is far enough into the oocyte for injection and further movement should be minimal. Inject, wait for a moment, and then withdraw the needle gently, ensuring that the oocyte does not get moved and its membrane become ripped.
9. Once the set of oocytes has been injected add 1X MBS, 3% ficoll to the dish, pipette the oocytes up gently and then place them in the same medium in a fresh Petri dish. Take care during this operation as, before the injection wounds heal, the oocytes are very sensitive to surface tension and will burst if they come into contact with the liquid's surface.

10. Incubate the oocytes at 18–21°C.
11. After 1 to 2 h replace the medium with 1X MBS and remove any unhealthy oocytes.
12. Culture as above for the appropriate period.

4. Notes

1. Care must be taken when preparing MBS, as the stock solutions will precipitate if they are mixed at high concentrations. In addition, the stocks themselves may, on occasion, precipitate, so examine these carefully before use. Some methods suggest storing these stocks frozen, we find this to be unnecessary and it can exacerbate the precipitation problem.
2. The choice of frog supply is extremely important. Until recently, we used wild-caught frogs for oocyte experiments, but preferred lab-bred ones for producing large numbers of healthy embryos. Recently, however, the quality of oocytes from wild-caught frogs seems to have decreased and we have started to use lab-bred frogs as a source of oocytes with great success, their added cost is currently good value. A number of suppliers offer suitable frogs, we have had excellent service from Nasco (Fort Atkinson, WI) (http://www.nascofa.com/Science/Xenopus.html). The conditions under which the frogs are kept are also critical, especially when the frogs are to be used often: If the water in your area is heavily chlorinated, it should be allowed to stand for 24 h before use to allow the chlorine to evaporate. The temperature should be between 18°C and 21°C with a light cycle of 14 h light, 10 h dark. The frogs should be fed twice weekly, we base their diet around frog pellets but occasionally add liver or heart, and uneaten food is removed after 1 h and the tanks are cleaned the next morning. Choosing between a manual and automated tank system is also necessary. We have experience using both, and now feel that should an automated system fail for any reason, the consequences are such (we ended up replacing all our stock) that it is better to use a manual one if the technical assistance for twice weekly cleaning is available.
3. Ethyl-*m*-aminobenzoate (also known as tricaine or MS222) is toxic and a carcinogen, wear gloves when handling it or a frog that has been anesthetized with it. Prolonged exposure of females to this chemical (>1 h) will also cause oocytes to mature.
4. In order for the oocytes to survive in culture, they must be divided rapidly into small clumps to provide oxygenation, and kept clean. Damaged oocytes and embryos seem to damage others around them.
5. Two methods are available to inject oocyte nuclei (germinal vesicles). The oocytes may be centrifuged to bring the nuclei to the top of the animal pole, where they become visible, or the injection may be carried out "blind." In the latter case, it is necessary to practice and ensure that your technique results in consistent delivery of injected material to the nucleus. Load a needle with a solution of a dye such as Nile Blue and inject oocytes, aiming for the center of the animal half. Carefully rip the oocytes apart using watchmaker's forceps and examine where the dye has gone; with practice you will find that it is nuclear in >90% of oocytes. To visualize the nuclei by centrifugation, place the oocytes animal pole

up on injection grids with the Petri dishes full of 1X MBS. Centrifuge in the bottom of 1-L buckets at 500 g for 10 min at 20°C. The nucleus is clearly visible, as it has displaced the pigment granules at the top of the animal pole; carry out the injections within the next 30–60 min because the nuclei do sink back to their original positions.

6. Embryo injection is in some ways easier than that of oocytes; physically, the membrane is easier to penetrate, and if targeting the injection is necessary, at least the target is visible. However, these are balanced by the need for injection during a short period, thus, it is essential to be well organized and prepared for injection before the embryos reach the appropriate stage. During injection at the one-cell stage, some embryos may be very hard to penetrate, these are unfertilized and should be ignored. Do not risk breaking a needle and having to recalibrate in mid-experiment by trying to stab these. Once injected, the embryos should be transferred to 0.1X MBS when they have developed to about stage 6.

7. Modern needle pullers will produce extremely consistent needles for micronjection. It is necessary to strike a balance in needle thickness to accommodate having the necessary stiffness to penetrate oocyte and embryo membranes, minimal injection wounding, and no blockages. Most people seem to find a setup that produces this balance by trial and error. The exact size will depend on your experiment, but as a guide, we use slightly thicker needles for oocyte injection than for one- and two-cell embryos, and then thinner ones again for later-stage embryos.

8. Although DNA prepared by a variety of methods may be used for injection, we have found that survival rates beyond gastrulation are markedly increased when the DNA is purified over two cesium chloride gradients and then using glass beads (e.g., geneclean or bandprep kits). DNA so prepared gives survival rates of >95% when 100–200 pg of reporter constructs are injected.

Acknowledgments

Many thanks to Fiona Myers for comments on the manuscript. Work in the Molecular Embryology Laboratory at Portsmouth is supported by the Wellcome Trust, BBSRC, and Cancer Research Campaign.

References

1. Colman, A. (1984) Transcription and Translation: A Practical Approach (Hames, B. D. and Higgins, S. J., eds.), IRL Press, Oxford, pp. 271–302.
2. Soreq, H. and Seidman, S. (1992) *Xenopus* oocyte microinjection—from gene to protein. *Methods Enzymol.* **207,** 225–265.
3. Zhao, J., Kung, H. F., and Manne, V. (1994) Farnesylation of P21 Ras proteins in *Xenopus* oocytes. *Cell. Mol. Biol. Res.* **40,** 313–321.
4. Lam, A., Kloss, J., Fuller, F., Cordell, B., and Ponte, P. A. (1992) Expression cloning of neurotrophic factors using *Xenopus* oocytes. *J. Neurosci. Res.* **32,** 43–50.
5. Yao, S. Y. M., Muzyka, W. R., Elliott, J. F., Cheeseman, C. I., and Young, J. D. (1998) Cloning and functional expression of a cDNA from rat jejunal epithelium

encoding a protein (4F2hc) with system y(+)L amino acid transport activity. *Biochem. J.* **330,** 745–752.
6. Castagna, M., Shayakul, C., Trotti, D., SacchV. F., Harvey, W. R., and Hediger, M. A. (1998) Cloning and characterization of a potassium-coupled amino acid transporter. *Proc. Natl. Acad. Sci. USA* **95,** 5395–5400.
7. Lan, L., Bawden, M. J., Auld, A. M., and Barritt, G. J. (1996) Expression of Drosophila Trpl cRNA in *Xenopus laevis* oocytes leads to the appearance of a calcium channel activated by calcium and calmodulin. *Biochem. J.* **316,** 793–803.
8. Rettinger, J. (1996) Characteristics of sodium/potassium-ATPase mediated proton current in sodium free and potassium free extracellular solutions—indications for kinetic similarities between H+/K+-ATPase and NA+/K+-ATPase. *Biochim. Biophys. Acta.* **1282,** 207–215.
9. Karbach, D., Staub, M., Wood, P. G., and Passow, H. (1998) Effect of site-directed mutagenesis of the agrinine residues 509 and 748 on mouse band 3-mediated anion transport. *Biochim. Biophys. Acta.* **1371,** 114–122.
10. Marshall, B. A., Murata, H., Hresko, R. C., and Mueckler, M. (1993) Domains that confer intracellular sequestration of the Glut4 glucose transpoter in *Xenopus* oocytes. *J. Biol. Chem.* **268,** 26,193–26,199.
11. Lee, D. H., Bennet, S., and Pedersen, K. (1995) Evidence against a potential endoplasmic reticulum transmembrane domain of 27K Zein expressed in *Xenopus* oocytes. *Protein Eng.* **8,** 91–96.
12. Kass, S. U., Landsberger, N., and Wolffe, A. P. (1997) DNA methylation directs a time dependent repression of transcription initiation. *Curr. Biol.* **7,** 157–165.
13. Leonard, M. W. and Patient, R. K. (1991) Evidence for torsional stress in transcriptionally activated chromatin. *Mol. Cell. Biol.* **11,** 6128–6138.
14. Wong, J. M., Li, Q., Levi, B. Z., Shi, Y. B., and Wolffe, A. P. (1997) Structural and functional features of a specific nucleosome containing a recognition element for the thyroid hormone receptor *EMBO J.* **16,** 7130–7145.
15. Wong, J. M., Patterton, D., Imhof, A., Gushin, D., Shi, Y. B., and Wolffe, A. P. (1998) Distinct requirements for chromatin assembly in transcriptional repression by thyroid hormone receptor and histone deacetylase. *EMBO J.* **17,** 520–534.
16. Moreau, N., Laine, M. C., Billoud, B., and Angelier, N. (1994) Transcription of amphibian lampbrush chromosomes is disturbed by microinjection of HSP70 monoclonal antibodies. *Exp. Cell Res.* 211, 108–114.
17. Roy, L. M., Haccard, O., Izumi, T., Lattes, B. G., Lewellyn, A. L., and Maller, J. L. (1996) Mos protooncogene function during oocyte maturation in *Xenopus*. *Oncogene* **12,** 2203–2211.
18. Wylie, C., Kofron, M., Payne, C., Anderson, R., Hosobuchi, M., Joseph, E., and Heasman, J. (1996) Maternal beta-catenin establishes a dorsal signal in early *Xenopus* embryos. *Development* **122,** 2987–2996.
19. Matsumoto, K., Meric, F., and Wolffe, A. P. (1996) Translational repression dependent on the interaction of the *Xenopus* Y-box protein FRGY2 with messenger RNA—role of the cold shock domain, tail domain, and selective RNA sequence recognition. *J. Biol. Chem.* **271,** 22,706–22,712.

20. Brewer, A. C., Guille, M. J., Fear, D. J., Partington, G. A., and Patient, R. K. (1995) Nuclear translocation of a maternal CCAAT factor at the start of gastrulation activates *Xenopus* GATA–2 transcription. *EMBO J.* **14(4),** 757–766.
21. Leibham, D., Wong, M., Cheng, T., Schroeder, S., Weil, P. A., Olson, E. N., et al. (1994) Binding of TFIID and MEF2 to the TATA element activates transcription of the *Xenopus* MyoDa promoter. *Mol. Cell. Biol.* **14,** 686–699.
22. Ovsenek, N., Zorn, A.M., and Krieg, P.A. (1992) A maternal factor, OZ-1, activates embryonic transcription of the *Xenopus laevis* GS17 gene. *Development* **115,** 649–655.
23. Santoro, B., Degregorio, E., Caffarelli, E., and Bozzoni, I. (1994) RNA-protein interactions in the nuclei of *Xenopus* oocytes—complex formation and processing activity on the regulatory intron of ribosomal protein gene L1. *Mol. Cell. Biol.* **14,** 6975–6982.
24. Tigyi, G. and Parker, I. (1991) Micrinjection into *Xenopus* laevis oocytes—a precise semiautomatic instrument and optimal parameters for injection of messenger RNAs. *J. Biochem. Biophys. Methods* **22,** 243–252.
25. Heasman, J. (1997) Patterning the *Xenopus* blastula. *Development* **124,** 4179–4191.
26. Hopwood, N. D. and Gurdon, J. B. (1990) Activation of muscle genes without myogenesis by ectiopic expression of MyoD in frog embryo cells. *Nature* **347,** 197–200.
27. Amaya, E., Musci, T. J., and Kirschner, M. W. (1991) Expression of a dominant negative mutant of the FGF receptor disrupts mesoderm formation in *Xenopus* embryos. *Cell* **66,** 257–270.
28. Suzuki, A., Kaneko, E., Ueno, N., and Hemmati-Brivanlou, A. (1997) Regulation of epidermal induction by BMP2 and BMP7 signaling. *Dev. Biol.* **189,** 112–122.
29. Joseph, E. M. and Melton, D. A. (1998) Mutant Vg1 ligands disrupt endoderm and mesoderm formation in *Xenopus* embryos. *Development* **125,** 2677–2685.
30. Piccolo, S., Agius, E., Lu, B., Goodman, S., Dale, L., and DeRobertis, E. (1997) Cleavage of chordin by Xolloid metalloprotease suggests a role for proteolytic processing in the regulation of Spemann organizer activity. *Cell* **91,** 407–416.
31. Conlon, F. L., Sedgwick, S. G., Weston, K. M., and Smith, J. C. (1996) Inhibition of Xbra transcription activation causes defects in mesodermal patterning and reveals autoregulation of Xbra in dorsal mesoderm. *Development* **122,** 2427–2435.

11

Microinjection into Zebrafish Embryos

Qiling Xu

1. Introduction

Microinjection remains the most popular and effective of the methods to introduce DNA, RNA, and proteins into fertilized zebrafish eggs. The method is simple and reliable. A microinjection pipet is filled with the DNA or RNA solution and attached to an apparatus that forces the solution out of the pipet with air pressure. A small amount of solution is then expelled into the cytoplasm of the embryo before withdrawing the pipet, and the injected embryos are incubated to develop further. Once inside the cells, the foreign DNA or RNA is transcribed and/or translated within the developing embryos and the functional roles of their protein products can be evaluated by morphological, physiological or molecular changes. Thus, microinjection has been widely used for generating transgenic fish *(1–3)*, analyzing gene function by overexpression of DNA or RNA *(4–6)* and mapping cell fate in early blastula embryos *(7,8)*.

The injected DNA first undergoes amplification and then gets integrated in the chromosomes, although there is some episomal DNA. The delay in integration and rapid cell division of the early zebrafish embryos leads to only a small fraction of the cells within the embryo inheriting the foreign DNA. Hence, the expression of the transgene is highly mosaic, and there is low-efficiency germ-line transmission of the transgenes by microinjection. To achieve more efficient germ-line transmission, other methods of gene transfer into the zebrafish embryos should be explored (e.g., retroviral infection) *(9,10)*. The distribution of injected RNA is more widespread than that of DNA, and overexpression as well as ectopic expression occurs very early in the developing embryos. However, the expression is not sustained during development because of degradation of the injected RNA, and how long the protein products will last depends on the stability of the expressed protein.

From: *Methods in Molecular Biology, Vol. 127: Molecular Methods in Developmental Biology: Xenopus and Zebrafish* Edited by: M. Guille © Humana Press Inc., Totowa, NJ

RNA injection is used mainly to analyze gene function in early events of embryonic development.

This chapter describes the equipment required for microinjection, the assembly of these components into a microinjection system, and the process of microinjection itself. The basic setup includes a pressure regulator, a micropipet holder, and a micromanipulator. The holder for the injection pipet is mounted on a micromanipulator and connected to the pressure regulator. The pressure is controlled with an automatic system and discharges are activated with a foot paddle. It is important to regulate the pressure, both up and down, smoothly while the pipet tip is still inside the cell. Use of an automatic injection system ensures controlled regulation of pressure and, hence, the reproducibility of each microinjection. It also simplifies and thus speeds up the injection process. In the simplest method, microinjection of one-cell stage zebrafish embryos can be carried out satisfactorily by holding the micropipet manually (with steady hands). However, a more refined and precise injection can be achieved using a micropipet holder and a micromanipulator. Detailed procedures of how to prepare embryos, DNA, and RNA for microinjection are described elsewhere in this book. In brief, the DNA should be linearized or isolated from the vector sequences and is dissolved at a concentration of 0.1 mg/mL in distilled water containing 0.2 M KCl. Higher concentrations will lead to higher lethality and nonspecific deformity because of the toxic effects of the DNA. In vitro transcribed RNA should be treated with RNase-free DNase to remove the template, extracted with phenol–chloroform and then purified on an RNase-free microspin column (e.g., S-400 HR column, Amersham Pharmacia Biotech, Buchinghashire, UK). The optimal RNA concentration is dependent on the nature of the encoded protein; for example, embryos have very low tolerance of ectopically expressed transcription factors and can have different responses to different concentrations of signaling molecules. In general, RNA is dissolved in pure water at a concentration between 0.01 and 1 mg/mL for microinjection. The appropriate injection volume is about 10–20% of that of the cytoplasmic volume, which is about 1 nL for the one-cell stage. Larger volumes can burst cells or cause nonspecific deformity of the embryos.

2. Materials

1. Dissecting microscope (e.g., Nikon SMX10TD Nikon UK Limited, Surrey, UK, Leica Wild M8, Leica UK Limited, Milton Keynes, UK).
2. Micromanipulator: This is used to position the tip of an injection pipet at a precise location within the embryo. Commonly used micromanipulators are the M type supplied by Leitz (Wetzlar, Germany) and the micromanipulation system by Narishige (Tokyo, Japan). These provide fine control of three-dimensional movement. The Leitz system requires a purpose-built base plate, and the Narishige system can be immobilized on a metal plate through a magnetic stand.

3. Micropipet holder: The holder for the microinjection pipet allows passing of the pressure through its connection to a pressure regulator via a polyethylene tube. Pressure-fitting micropipet holders are available from Narishige or World Precision Instruments (Sarasota, NY).
4. Microinjection pipets: Microinjection pipets are prepared from standard-walled glass capillaries with an inner filament (e.g., Clark Electromedical Instruments, Reading, UK, catalog number GC100F-15). Capillaries with an internal filament allow the pipet to be backfilled, by capillary action, from the butt end to the injection tip. Microinjection pipets are drawn using a pipet/needle puller (e.g., David Kopf [Tujunga, USA], Narishige PB-7).
5. Microinjection chamber: The injection chamber is used to hold embryos during microinjection. Embryos can be held in a simple trough, generated by placing a microscope slide in a Petri dish or in wedge-shaped troughs or wells in 1.5% agarose made with a plastic mold (*see* **ref. 11**). The agarose mount is useful because the pipet tips will not break if they accidentally touch the surface.
6. Pressure regulator: An automatic injection system such as Inject+matic (Geneva, Switzerland) or Picospritzer (General Valve Inc., USA) is essential to deliver constant pressure without hysteresis and to control the duration of injection. The Inject+matic uses a motor to generate pressure and can generate suction so that it can be used to fill the micropipet (although I prefer to backfill as described below), whereas the Picospritzer uses compressed nitrogen or air to deliver a sharp pulse of pressure.
7. Glass Pasteur pipets: To use for transferring embryos, an opening with the size fitting an embryo is cut with a diamond pen and fire-polished.
8. Microscope slides.
9. Petri dishes.
10. Incubator (28.5°C).
11. Watchmaker's forceps (Dumont, No. 5, Switzerland).
12. Microloader pipet tips (Eppendorf, Germany).

3. Methods
3.1. Assembly of the Microinjection System

A typical microinjection system is shown in **Fig. 1**.

3.1.1. The Microinjection Pipet

1. Draw a rather steep taper and very sharp tip on the injection pipets with a standard puller. When determining the settings, check the tip of the pipet under a dissecting microscope. It should be sharp with the tip closed. Prepare several pipets and store them in a dust-free container.
2. Immediately before the injection, break off the tip of the microinjection pipet with a pair of forceps or a razor blade. Backfill the injection pipet by placing a drop of solution containing DNA, RNA, or dye on the butt end. This can be done using a standard pipet tip or a microloader pipet with a tip that fits inside the

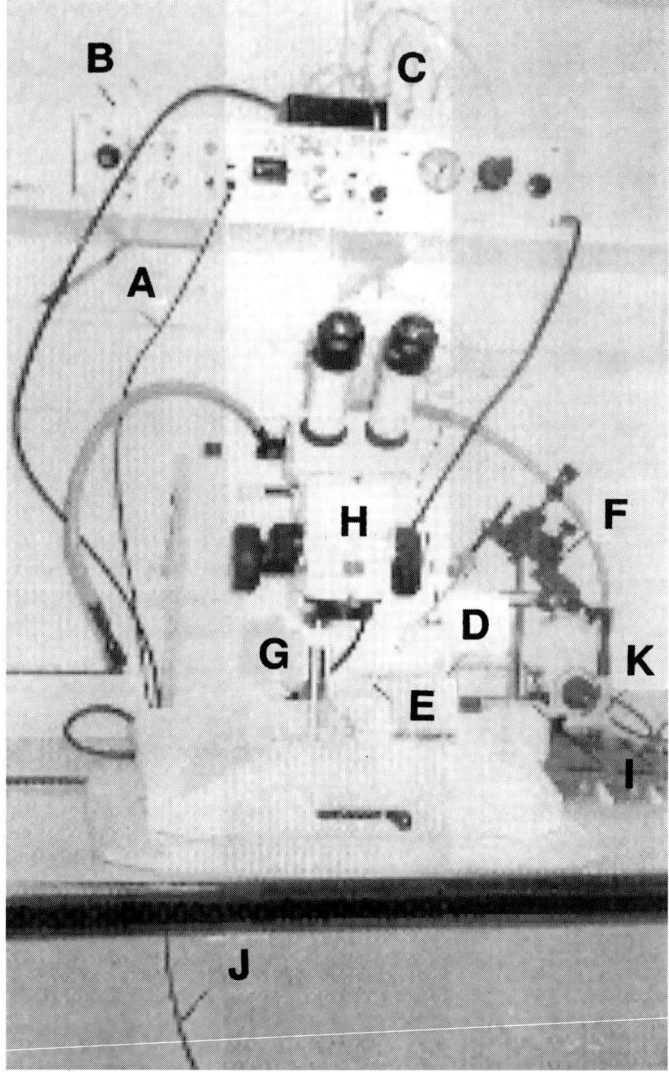

Fig. 1. Typical setup for microinjection of one- to two-cell stage zebrafish embryos: (A) Connection to a compressed air cylinder; (B) Automatic pressure regulator (Picospritzer); (C) connection to an airtight pipet holder; (D) Micropipet holder; (E) Microinjection pipet; (F) Micromanipulator; (G) Injection chamber; (H) Dissecting microscope; (I) Magnetic stand and base plate; (J) Connection to a foot paddle; (K) Light source for microscope.

capillary tube. The inner filament will draw the liquid to the tip. Insert the filled pipet into a pressure fitting holder and tighten the screw to secure the pipet. Mount the holder on a micromanipulator and make sure that there is nothing in the way of the injection pipet.

3. Calibrate the volume injected by this needle by placing a graticule under a small Petri dish containing mineral oil, deliver a few separate injections into the oil, and measure the diameter of the resulting droplets. Convert this into volume and alter the settings (duration of pressure pulse or its pressure) on the microinjector to deliver the required volume (*see* **Note 4**); recheck again as above.

3.1.2. The Microinjection Chamber

There are two types of injection chambers. One type is very simple and needs only a Petri dish and a microscope slide. No preparation is required and it enables easy mounting and dismounting of embryos before and after the injection, thus speeding up the injection process. The agarose mold type *(11)* provides a safety cushion to prevent accidental breakage of the pipet tips, and the agarose cushion is necessary if dechorionated embryos are used for injection. However, accidental breakage is relatively rare and embryos can be easily injected in their chorions. Furthermore, special care is needed to culture the dechorionated embryos, as they are more vulnerable to physical damage and infection. The following only describes the simple method because it serves its purpose well.

1. Place a microscope slide in a 100-mm Petri dish and add a few drops of embryo media to create a thin film of liquid to hold the slide in place.
2. Tilt the Petri dish to drain the liquid and remove any excess liquid.

3.2. Microinjection of Zebrafish Embryos

3.2.1. Preparing Embryos

1. Collect embryos 20–30 min after natural spawning. Check the embryos under a dissecting microscope and select one- to two-cell stage embryos for microinjection (*see* **Note 5**). It is feasible to inject both cells of two-cell stage embryos.
2. Keep the rest of embryos in an 18°C incubator for later use. Embryos can tolerate 18°C for about 1 h and this lower temperature slows down the cell division rate, hence allowing more time to inject embryos at one to two-cell stages.
3. Use a diamond pen to cut a glass Pasteur pipet and fire-polish the edge. The point should just fit the size of an embryo. Squirt the embryos one by one along the edge of the slide and align them neatly (*see* **Fig. 2**). Tilt the dish to drain the liquid and remove as much liquid as possible, but leave enough to coat the embryos and make sure they are moist during injection. One slide length can accommodate 40–50 embryos; this is a manageable number in one injection batch.
4. Use forceps to orientate the embryos with the cells perpendicular to the trough or in a dorsal orientation. This step is optional but helps the beginner.

3.2.2. Microinjection

1. Under low magnification, use the manipulator to bring the pipet tip close to the intended embryo (usually start from one end) and use higher magnification

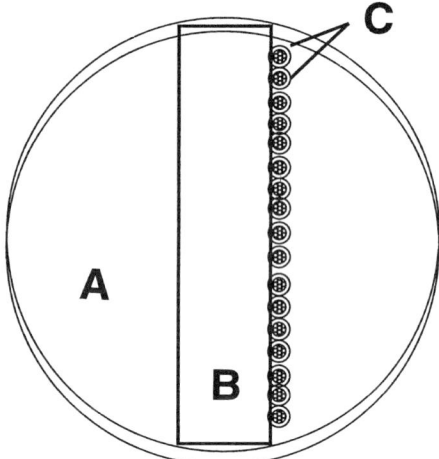

Fig. 2. Illustration of the microinjection chamber: (**A**) Petri dish; (**B**) microscope slide; (**C**) zebrafish embryos.

(35× to 40×) to locate the tip at the surface of chorion. Focus on the cells to be injected. Use light transmission between dark- and bright-field illumination to obtain a clear, contrasting image of the embryos.
2. Make a pressure discharge using the foot switch to ensure that the pipet is not blocked. A small drop should appear at the needle tip after the discharge. If not, adjust the injection pressure and duration or increase the tip size. Sometimes, simply dipping the tip in the liquid can remove the blockage.
3. Using the controls on the micromanipulator, drive the tip of the pipet through the chorion and, in a single smooth move, into the cytoplasm of the embryo. A steep angle (around 45°) between the pipet tip and cells facilitates easier penetration and prevents the embryos from slipping upon the impact of the tip after it goes through the chorion. Enter directly into the embryonic cell rather than through the yolk because the yolk is sticky and more likely to block the tip.
4. Make a discharge to deliver the liquid. The injected solution (*see* **Notes 1** to **4**) is visible within the cytoplasm, as its constituents differ greatly than those found within the cell. Failure to see the discharge of the liquid may indicate two possibilities. One is a blockage of the injection pipet. Try to unblock it by increasing the injection pressure and/or duration. Alternatively, enlarge the tip; to do this, withdraw the pipet first and nip off the tip slightly either by gently touching the plastic surface with it or using a pair of forceps. If the tip remains blocked after these treatments, change the pipet. The second possibility is that the tip failed to go through the cell membrane. There are two things that could contribute to such failure. The pipet tip may be too blunt or the angle between the tip and cells may be too flat. Adjust accordingly and repeat **step 3**.

5. Withdraw the tip from the embryo by slowly lifting the pipet. The embryo will be held down by the surface tension. Two things can help in withdrawing the pipet smoothly. First, have as little liquid as possible in the injection chamber to hold embryos relatively tightly by surface tension. Second, have a relatively long tapered tip so very little resistance occurs while the tip is in the cell membrane and chorion.
6. Move the dish slightly to locate the next embryo, and inject it. Inject all of the embryos in the batch.
7. Tilt the chamber and add embryo media. Embryos should slide down freely to the lower part of the dish. Transfer them to a dish with embryo media and incubate at 28.5°C.

3.3. Postinjection Care of Embryos and Preparation for Analyses

1. Check the embryos a few hours after injection and remove any dead or grossly abnormal embryos.
2. For enzymatic detection, *in situ* hybridization, or immunocytochemical analyses, injected embryos before the 15 somite stage can be fixed in their chorions and then dechorionated after fixation. Embryos older than that are dechorionated prior to fixation.
3. For embryos subsequently used for cell transplantation or close morphological examination, dechorionate them 1 h after injection and incubate them in a 1.5% agarose-coated dish.
4. To dechorionate the embryos manually use two pairs of watchmaker's forceps (Dumont no. 5). First, grip the chorion firmly with one pair, ensuring that you do not contact the embryo. With the second pair, take a grip next to the first and move the two pairs apart slowly. Allow the embryo to be released slowly, as there is a change in pressure during this process and rapid release often damages the embryo.

4. Notes

1. The quality of the DNA or RNA solution is crucial to ensure correct assessment of the effects on the development of zebrafish embryos. Impurities in the injection solution can cause nonspecific abnormalities or even death and contribute to false conclusions. The best way to purify DNA or RNA is to use a microspin column (Pharmacia) and use the highest grade water (e.g., W-4502, Sigma, Poole, UK) to dilute it to the desired concentration. EDTA will kill zebrafish embryos.
2. High DNA concentration (>0.1 mg/mL in a 200-pL injection) have a toxic effect on the embryos. The results range from widespread nonspecific abnormality to early lethality. However, there is a balance that must be determined empirically to achieve widespread expression without toxicity.
3. Initial experiments should be carried out to find out the appropriate concentration of the RNA solution. Tolerance of embryos to RNA samples depends on the nature of the protein it encodes.
4. Too large an injection volume will either burst the cell membrane or disturb the cytoplasmic components in the cells leading to abnormal development. One

deformity often associated with injecting too large a volume is bent tails. A volume of 200 pL is ideal.
5. The first cleavage time of zebrafish embryos is around 10–30 min after fertilization. Microinjection at the one-cell stage has to be carried out within this time, although keeping embryos at 18°C will slow down cell division. It is, therefore, advisable to prepare the injection pipets and solutions in advance.

References

1. Stuart, G. W., McMurray, J. V., and Westerfield, M. (1988) Replication, integration and stable germ-line transmission of foreign sequences injected into early zebrafish embryos. *Development* **103,** 403–412.
2. Stuart, G. W., Vielkind, J. R., McMurray, J. V., and Westerfield, M. (1990) Stable lines of transgenic zebrafish exhibit reproducible patterns of transgene expression. *Development* **109,** 557–584.
3. Culp, P., Nusslein-Volhard, C., and Hopkins, N. (1991) High frequency germ-line transmission of plasmid DNA sequences injected into fertilised zebrafish eggs. *Proc. Natl. Acad. Sci. USA* **88,** 7953–7957.
4. Kelly, G. M., Greenstein, P., Erezyilmaz, D. F., and Moon, R. T. (1995) Zebrafish Wnt8 and Wnt 8b share a common activity but are involved in distinct developmental pathways. *Development* **121,** 1787–1799.
5. Xu, Q. L., Alldus, G., Holder, N., and Wilkinson, D. G. (1995) Expression of truncated Sek-1 receptor tyrosine kinase disrupts the segmental restriction of gene-expression in the Xenopus and zebrafish hindbrain. *Development* **121,** 4005–4016.
6. Long, Q. M., Meng, A. M., Wang, H., Jessen, J. R., Farrell, M. J., and Lin, S. (1997) GATA-1 expression pattern can be recapitulated in living transgenic zebrafish using GFP reporter gene. *Development* **124,** 4105–4111.
7. Strehlow, D., Heinrich, G., and Gilbert, W. (1994) The fates of the blastomeres of the 16-cell zebrafish embryo. *Development* **120,** 1791–1798.
8. Helde, K. A., Wilson, E. T., Cretekos, C. J., and Grunwald, D. J. (1994) Contribution of early cells to the fate map of the zebrafish gastrula. *Science* **265,** 517–520.
9. Lin, S., Gaiano, N., Culp, P., Burns, J. C., Friedmann, T., Yee, J.-K., et al. (1994) Integration and germ-line transmission of a pseudotyped retroviral vector in zebrafish. *Science* **265,** 666–669.
10. Gaiano, N., Allende, M., Amsterdam, A., Kawakami, K., and Hopkins, N. (1996) Highly efficient germ-line transmission of proviral insertions in zebrafish. *Proc. Natl. Acad. Sci. USA* **93,** 7777–7782.
11. Westerfield, M. (1995) *The Zebrafish Book, A Guide for the Laboratory Use of Zebrafish*, 3rd ed., University of Oregon Press, Eugene, OR.

12

Expression from DNA Injected into *Xenopus* Embryos

Ondine Cleaver and Paul A. Krieg

1. Introduction

Expression of genes by introduction of DNA into developing embryos can be used to analyze promoter elements that are developmentally regulated or to ectopically express proteins to test their function during development. Because of its large size and resilience to manipulation, the *Xenopus* embryo is particularly suitable for studies utilizing microinjection, and there is a large body of literature describing successful promoter analysis and ectopic expression experiments. In this chapter, we will outline the fundamental methods and tools for expression of injected DNA in the early embryos of *Xenopus*. This chapter will not, however, attempt to review the recently developed *Xenopus* transgenesis procedures *(1,2)*. Although these transgenic methods offer significant advantages over the transient expression methods that we will describe below, the method is still under development and few examples of promoter analysis or of ectopic gene expression studies are yet available. Meanwhile, DNA expression from injected templates provides an efficient and simple method for gene analysis in the embryo.

The first DNA injection experiments were carried out using *Xenopus* oocytes. Rather than embryos, Colman *(3)* injected synthetic polynucleotides into both *Xenopus* oocytes and eggs and observed a significant increase in RNA synthesis following the injection, suggesting that the DNA was being transcribed. Mertz and Gurdon *(4)* extended this use of the *Xenopus* oocyte as a transcriptional vehicle and found increased transcription from injection of SV 40 DNA directly into the germinal vesicle (nucleus). Since then, a large number of different DNA constructions have been injected into oocytes, including DNAs encoding viral genes, histone genes, and ribosomal RNA genes *(5–15)*.

From: *Methods in Molecular Biology, Vol. 127: Molecular Methods in Developmental Biology: Xenopus and Zebrafish* Edited by: M. Guille © Humana Press Inc., Totowa, NJ

Experiments using the injection of plasmid DNA in fertilized *Xenopus* eggs or in cleavage-stage *Xenopus* embryos have also been carried out to assess the regulation of gene expression and plasmid DNA replication during development and to analyze gene function. Early experiments involving injection of DNA into *Xenopus* embryos included the injection of sea urchin histone, rabbit β-globin, and *Drosophila* ADH genes into fertilized *Xenopus* eggs *(16–19)*.

1.1. General Properties of DNA Expression

In many cases, the DNA constructions injected into *Xenopus* embryos contain a constitutive promoter used to drive gene expression. However, despite the presence of the constitutive promoter, expression from these constructions does not occur until the mid-blastula transition (MBT) when transcription is activated in the *Xenopus* embryo *(20,21)*. The MBT occurs approximately 8 h after fertilization, when the embryo consists of about 4000 cells. The vast majority of the injected DNA remains extrachromosomal and is incorporated into nucleuslike structures that resemble normal cell nuclei, including a nuclear membrane *(22–24)*. This unintegrated plasmid DNA remains stable throughout early development and is easily detectable until, and including, the late tailbud stages, at levels close to the amount originally injected. In general, very low levels of injected DNA do not replicate and very high levels may lead to toxic effects (see below for more details). Finally, expression from the injected DNA is mosaic, meaning that only a subset of the cells containing the plasmid DNA actually transcribe the template sequences. At present, the molecular mechanism responsible is not understood.

1.2. Analysis of Promoter Activity

There are a number of reported examples where genes appear to exhibit correct temporal and tissue-specific expression when introduced into the *Xenopus* embryo. Examples of correct temporal expression include GS17 *(25,26)*, cardiac actin *(27,28)*, keratin *(29)*, hsp 70 *(30)*, cytoskeletal actin *(31)*, Xnf7 *(32)*, and hepatocyte growth factor (HGF) *(33)*. Examples of correct spatial expression include α-actin *(34)*, keratin *(29)*, XmyoDa *(35)*, Xnf7 *(32)* and HGF *(33)*. In general, these genes were expressed from plasmids that were not integrated into the chromosomal DNA but remained extrachromosomal. In addition, although largely correct tissue distribution was observed, expression remained clearly mosaic. Although there certainly have been cases where correct tissue specific expression has been observed, there are also many examples where repeated efforts have failed to demonstrate correct expression in the embryo. The reasons for the failure to observe tissue specific expression are not clear, because in at least some cases, the same promoter constructions that fail to show tissue-specific expression after DNA injection into embryos exhibit correct tissue-specific expression in transgenic animals *(1)*. This suggests that the limitations are not the result of the sequences themselves, but perhaps to secondary effects such as DNA conformation or methylation.

Promoter analysis can be carried out using a number of convenient plasmid constructions containing different reporter genes, including green fluorescent protein (GFP), β-galactosidase, luciferase and chloramphenicol acetyltransferase (CAT). The β-galactosidase and GFP reporters are particularly useful when assaying for tissue-specific expression from a promoter construction, because they allow direct visualization of the expressing cells. However, there are also good antibodies against CAT and GFP, which allow the expressing cells to be detected *in situ* following immunocytochemical detection. On the other hand, the CAT and luciferase reporters offer the advantage of straightforward quantitation assays that greatly facilitate analysis of the relative transcriptional activity of different promoter constructions. By comparison, it is relatively difficult to quantitate β-galactosidase and GFP expression levels. Some examples of the use of different reporter genes are contained in the following references: β-Galactosidase *(36–38)*, luciferase *(39–44)*, and CAT *(45–47)*.

1.3. Ectopic Expression of Genes in the Embryo

Another use of DNA injection is for the ectopic expression of experimental genes in the embryo. Depending on the way in which the procedure is carried out, the experimental gene has the potential to be (1) expressed in all cell types in the embryo, by the use of a constitutive promoter, or (2) expressed in a specific subset of embryonic tissues, either by using a tissue-specific promoter or by injecting a construct with a constitutive promoter into a specific blastomere with a restricted range of cell fates. In practice, the ectopic expression of gene products from injected DNA is particularly suitable for expression of proteins that do not act cell autonomously (e.g., secreted proteins or growth factors). For growth factors, the fact that expression is mosaic has little effect on interpretation of experimental results. Examples of successful ectopic growth factor expression from injected DNA constructs include Xwnt-8 *(48)*, BMP-4 *(49)*, and vascular endothelial growth factor (VEGF) *(50)*.

The choice of vector for expression is clearly important and each of the promoters in the expression plasmids described below are active constitutively and all are quite powerful. By placing the coding region of your gene of interest behind one of these promoters, it is possible to express quite high levels of mRNA. Because the promoters are constitutive, all different cell types receiving the injected construct will express the construct at significant levels. Note, however, that the mosaicism problem still restricts expression to only a proportion of the cells that actually contain the plasmid DNA. Although numerous

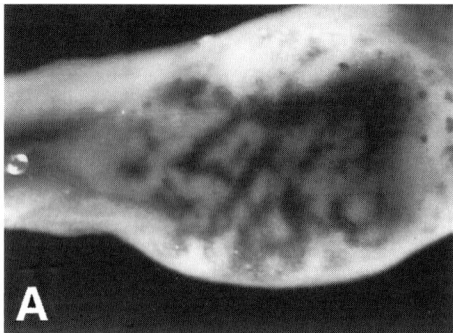 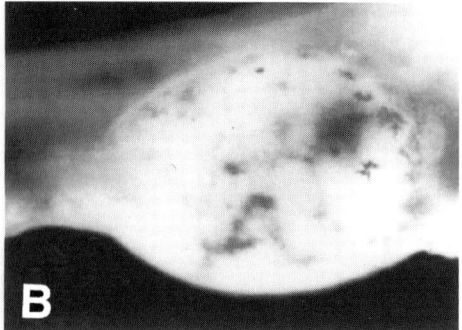

Fig. 1. Overexpression of VEGF using DNA injection. Ventral view of stage-43 embryos stained with benzidine to detect blood cells. The presence of blood is used to assay vascular structures. Anterior is to the right. (**A**) Embryo injected with 250 pg of pXeX–VEGF122 in a ventral vegetal blastomere at the 16-cell stage. Embryonic vascular structures are significantly enlarged and extended ectopically over the embryonic gut. (**B**) Control embryo showing normal vascular structures.

other vectors have been used in individual experiments for specific purposes, the following vectors have been used by many different laboratories and are known to be reliable and efficient:

Cytoskeletal actin promoter: Plasmids containing this promoter were originally constructed in Richard Harland's laboratory and have been used to express a number of different genes in the *Xenopus* embryo, including Xwnt-8 *(48)* and BMP-4 *(49)*. Transcription from this vector commences at the mid-blastula transition, peaks during gastrulation stages, and declines following neurulation.

Elongation factor 1-alpha promoter: pXeX is an efficient vector containing a compact version of the powerful *Xenopus* EF-1α promoter, a multiple cloning site, and a polyadenylation signal *(45)*. Expression from this promoter commences at the mid-blastula stage and gradually increases as development proceeds. We have used this promoter to express growth factor constructs in the *Xenopus* embryo, with good results *(50)*. An example of the use of this promoter to overexpress VEGF in the *Xenopus* embryo is illustrated in **Fig. 1**.

Cytomegalovirus promoter (CMV): Constructed by Dave Turner and Ralph Rupp, pCS2+ is an excellent expression vector containing the CMV promoter *(51)*. The simian CMV promoter drives high-level expression in most cell types, including mammalian, *Xenopus*, and zebrafish cells. It can be used either for in vivo expression following injection or transfection, or for in vitro mRNA synthesis. In addition to the very powerful CMV promoter, this construction contains a multiple-cloning site and the SV40 polyadenylation sequence. An example of the use of the CMV promoter in *Xenopus* is given by Kuhl et al. *(52)*.

1.4. DNA Versus mRNA Injection

Injection of DNA template for the ectopic expression of genes has several advantages relative to the injection of synthetic mRNA. It also has several disadvantages. Probably the major advantage is that the DNA template persists in the embryo throughout early embryonic development (at least until the early tadpole stages). Therefore, it is possible to ectopically express experimental sequences quite late in development, long after the majority of injected mRNA would be degraded. Levels of injected mRNA will generally decline rather rapidly and almost no transcripts are detected by the early to mid-tailbud stages *(53)*. Furthermore, because transcription from DNA templates does not commence until the mid-blastula stage, at the earliest, it is possible to avoid some potentially nonspecific effects of overexpression; for example, injection of mRNA encoding growth factors can have very dramatic effects on axis formation during early embryonic development that may mask more specific effects during later development. The use of a DNA vector makes it possible to delay the onset of expression of the growth factor until after the critical early stages of mesoderm induction. In fact, DNA constructions which include certain promoters can be used to ensure that transcription will begin at reasonably specific times during development; for example, the cytoskeletal actin promoter begins expression at mid-blastula but peaks around gastrulation *(49)*. As previously mentioned, the major disadvantage of expression from injected DNA templates is the fact that expression is mosaic. Recently, however, it has been reported that inclusion of viral inverted terminal repeats (ITR) sequences in the DNA expression vector can greatly reduce the degree of mosaicism and perhaps increase the likelihood of obtaining tissue-specific gene expression. So far, improved tissue specific expression has been demonstrated only for the cardiac α-actin promoter *(54)*. It will be interesting to determine whether the ITR-containing vector also results in tissue-specific expression of other embryonic promoters.

1.5. Transfection of DNA Constructions

As described below, the injection of individual blastomeres during early cleavage stages results in a relatively wide distribution of the injected material, and within this distribution, only a subset of cells will express the injected DNA construct (*see* **Fig. 2**). In birds and mammals, retrovirus-mediated transfer of DNA has allowed for very localized overexpression of certain DNA constructs *(55–57)*. However, this approach has not been useful for amphibian embryos. Transfection experiments in *Xenopus* embryos aimed at expressing exogenous DNA in certain neural tissues were originally carried out by Holt et al. *(58)*, who accomplished the transfection of a vector expressing luciferase cDNA into embryonic neurons, using several different methods. In one method,

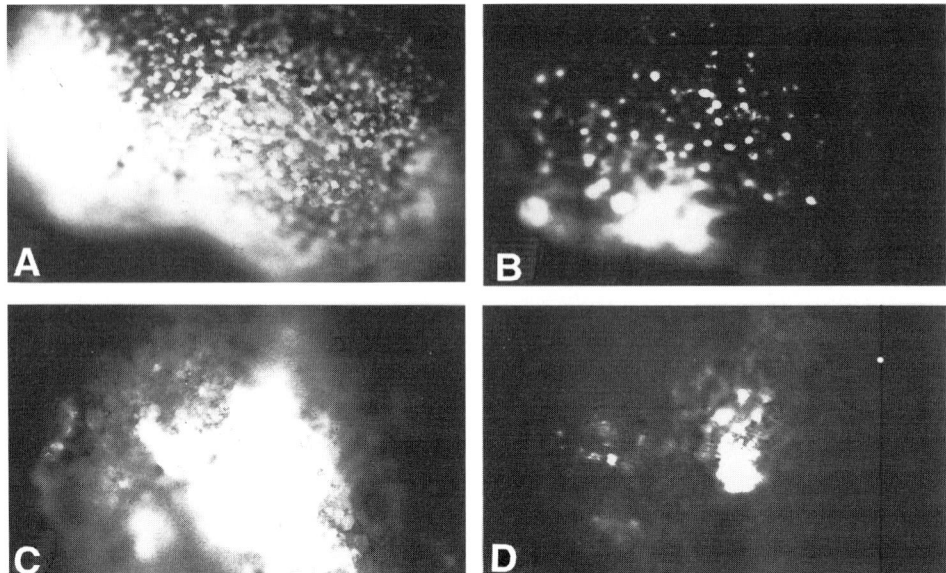

Fig. 2. Distribution of lineage tracer and GFP plasmid expression following injection or transfection. (**A**) Distribution of RDA in a stage-23 embryo, following injection of GFP plasmid and rhodamine mixture into the vegetal portion of a dorsal blastomere, CD1,2 of an eight-cell stage embryo. RDA fluorescence can be observed in a large portion of the embryo. (**B**) Distribution of GFP protein in the same embryo described in (**A**). Note that only a subset of the cells that show RDA fluorescence also show GFP fluorescence, indicating mosaic expression. (**C**) Distribution of RDA in a stage-23 embryo, following injection of transfection mixture under the epidermis on the flank of a stage-20 embryo. (**D**) Distribution of GFP protein in the same embryo as in (**C**). Note, once again, that only a subset of the cells which show RDA fluorescence also show GFP fluorescence.

excised tissue fragments from stage 20–24 *Xenopus* embryos were incubated directly in a solution containing the DNA construct and Lipofectin, a synthetic cationic lipid preparation. The transfected tissues were then reimplanted into recipient embryos. In a variation of the method, the DNA–Lipofectin (Bethesda Research Laboratories, Gaithersburg, MD) mixture was injected directly into the lumen of the eye vesicle of stage 20–24 embryos.

We have also carried out experiments based on the Holt laboratory protocol in an effort to achieve a more localized expression of DNA constructs in *Xenopus* embryos. We have used GFP as a reporter and targeted small groups of cells at various locations in the embryo by localized injection of a transfection mixture. If this localized transfection protocol can be made to work at high efficiency, it has the potential for expression of genes in specific embryonic locations. This method will be especially useful for the expression of molecules, such as growth factors, that do not function cell autonomously.

1.6. Conclusions

Overall, DNA injection into *Xenopus laevis* embryos provides an extremely useful and rapid assay of gene function. *Xenopus* embryos are highly resilient to the procedure and the subsequent disruption of their development, or of downstream gene expression, can yield important results in a short period of time. Despite the variability encountered due to mosaicism of expression, it is particularly suited for ectopic expression of non-cell autonomous gene products, such as growth factors. At this time, it is also much more straightforward and more reliable that the protocols for generation of transgenic animals, although we can expect the transgenic techniques to be constantly improved.

2. Materials

1. Microinjection apparatus: A thorough description of the microinjection apparatus is beyond the scope of this chapter, however, rather extensive descriptions are contained in **refs. *3*, *59*,** and ***60*** and Chapter 10 of this volume (*also see* **Note 1**).
2. Micropipet puller: Micropipets for injection (or injection needles) are produced using a Narishige PB-7 micropipet puller, although other models are just as effective. A two-stage pull technique allows for the production of micropipets with tips in the submicron to micron range in diameter, although for embryo injection, tips are generally produced with diameters between 10 and 50 µm. Needles should be pulled to acquire a long, fine taper of approximately 1 cm in length. Glass capillaries, such as the $3^1/_2$-in. or the 7-in. oocyte injector bores from Drummond Scientific Co. (Broomall, PA) (0.531 mm internal diameter), are recommended for microinjection.
3. DNA injection solution: The DNA to be injected is dissolved in sterile deionized water (*see* **Note 2**).
4. 2–3% Cysteine–HCl (pH 7.9–8.0).
5. Ficoll (Sigma, St. Louis, MO) Injection Buffer: Embryos are placed in this buffer while they are microinjected. There are several alternatives that give good results, but we usually place embryos in 100% Steinberg's solution, 3% Ficoll, and 25 µg/mL gentamycin about 5–15 min prior to the injection procedure. An alternative buffer that also works well consists of 0.5X Marc's modified Reinger's (MMR) solution, 3% Ficoll, and 25 µg/mL gentamycin.
6. Incubation buffer: After several hours in Ficoll injection buffer (2–12 h at 13°C), embryos should be transferred into incubation buffer, which consists of 20% Steinberg's solution with no Ficoll (or 0.2X MMR).
   Steinberg's solution:

Stock A (20X)
1.16 M NaCl
13.4 mM KCl
16.6 mM MgSO$_4$
6.7 mM Ca(NO$_3$)$_2$
pH to 7.4

Stock B (20X)
92.5 mM Tris base
0.08 N HCl

50 mL stock A + 50 mL stock B made to 1 L is 100% Steinberg's solution.
Marc's Modified Ringer's Solution (1X MMR): 100 mM NaCl, 2 mM KCl, 2 mM CaCl$_2$, 1 mM MgCl$_2$, 5 mM HEPES, pH 7.4.

7. Lineage tracer: DNA may be injected together with a lineage tracer to identify tissues that receive the injected material (*see* **Note 3**). Examples of useful lineage tracers are:
   FDA 5 mg/mL (fluorescein dextran amine, Molecular Probes, Eugene, OR).
   RDA 10 mg/mL (rhodamine dextran amine, Molecular Probes).
   β-Galactosidase mRNA 110 µg/mL.
   GFP (green fluorescent protein) mRNA 110 µg/mL *(61)*.
8. MEMPFA: 0.1 M 3-N-morpholino)propane sulfonic acid (MOPS) pH 7.4, 2 mM ethylene glycol-bis(β aminoethylether)-$N,N,N'N'$-tetraacetic acid (EGTA), 1 mM MgSO$_4$, 4% paraformaldehyde.
9. Paraformaldehyde stock solution: Heat water to 55–60°C, add paraformaldehyde in powder form, and add one to two drops of NaOH per 50 mL, otherwise the paraformaldehyde will not dissolve. Once paraformaldehyde has gone into solution, filter it to remove impurities.
10. β-Galactosidase staining solution (for 500 µL): 200 µL of 50 mM K$_4$Fe(CN)$_6$ and 200 µL of 50 mM K$_3$Fe(CN)$_6$ (store these two stock solutions in the dark, at room temperature), 1 µL of 1 M MgCl$_2$, 1 µL of 10% NP-40, 94 µL of PBS 1X, 4 µL of 100 mg/mL X-gal (warm the solution to 37°C before adding the X-gal to avoid X-gal precipitation).

3. Methods

3.1. General Protocol for Expression from Injected DNA

The following protocol has been used for the successful microinjection of DNA constructions into the *Xenopus* fertilized eggs and cleavage stage embryos.

1. Fertilized *Xenopus* eggs should be obtained as described by Newmeyer and Wilson *(62)* or Chapter 10 of this volume.
2. Before the first cleavage, remove the jelly coat from the fertilized eggs by incubating them in 2–3% cysteine-HCl (pH 7.9–8.0) for a few minutes with gentle swirling (you can see the jelly coat loosen and fall off the eggs) *(63)*.
3. When the jelly coats have been removed and the embryos can be seen to pack together closely, the fertilized eggs should be rinsed. Decant the cysteine solution and rinse the eggs gently multiple times with 20% Steinberg's over a period of approximately 10 min. All traces of cysteine must be removed to ensure continued viability. It is also important that the embryos do not come into contact

with the surface of the buffer or they are likely to rupture. Overall, this dejellying procedure is particularly important, as partially dejellied embryos are almost impossible to inject. On the other hand, embryos treated with cysteine-HCl for too long rupture easily and do not survive injection. It is useful to note that the cysteine solution becomes less active with time and that slightly longer incubations may be necessary for subsequent batches.

4. After the removal of the jelly coat, the embryos are incubated in sterile 20% Steinberg's solution either at 13°C or at room temperature, until the desired stage for injection, when they are transferred to the Ficoll Injection Buffer. Incubation at 13°C causes embryos to develop more slowly and allows for a longer window of time in which embryos at a specific stage can be injected. Injections can be performed from about the beginning of the first cleavage division during the 1-cell stage (about 90 min after fertilization at 13°C), until about the 64-cell stage (about 7 h after fertilization at 13°C). After the 64-cell stage, the individual blastomeres become rather difficult to inject using a normal injection apparatus, although later-stage cells can be injected using specialized equipment and smaller injection volumes.

5. For injections using a Nanojet variable automatic injector (Drummond Scientific, Co.): Load the micropipet with mineral oil (using metal needle provided with the Nanoject apparatus) and attach it to the Nanoject injector (*see* **Note 1**). Care should be taken when attaching the needle, because the plunger can be easily bent and damaged. Once attached, the needle should then be broken with forceps at the location where it just becomes flexible. The diameter of the resulting tip should be in the micron range (approximately 25 µm) and be rigid enough to penetrate the fertilization membrane of the embryo. A reticle micrometer can be used to accurately measure tip diameter.

6. Prepare the DNA injection solution containing lineage tracer, usually in a total volume of about 5 µL. Spin the DNA injection solution at maximum speed in a microfuge for 5 min and transfer the solution to a fresh tube. This spin reduces the likelihood that particulate matter in the injection solution will block the injection needle. When using the Nanoject apparatus, the DNA injection solution usually consists of the DNA construction of interest at a concentration of about 45 ng/µL in water, combined with a lineage tracer (*see* **Notes 3** and **4**). When using an injection volume of 4.6 nL (the standard injection volume of the Nanoject), this results in approximately 200 pg of plasmid DNA per injection. The DNA concentration can be up to four or five times higher, depending on the sequence used; however, toxicity may occur when using very high DNA concentrations, causing the injected cells to undergo necrosis and die. The presence of tracer molecules in the injection solution allows a clear identification of the tissues that received the injected material (*see* **Notes 2** and **5**).

7. Place 1–2 µL of the DNA injection solution on a clean piece of parafilm and fill the microinjection needle. Draw the DNA injection solution into the micropipet slowly, so that air bubbles do not develop in the oil around the injector plunger. If using the Nanoject, keeping the white gasket clean and the micropipet base blunt

(not chipped) significantly reduces the chance of air entering the oil. A few small bubbles will not cause any problems, but larger bubbles may create variability in the oil pressure and consequently in the volumes injected. At a DNA concentration of about 45 ng/µL, the 4.6-nL minimum injection volume permitted by the Nanoject system contains about 200 pg of DNA (*See* **Note 5** for the possible effects of injecting large amounts of DNA.)

8. Place about 100 embryos, at a developmental stage suitable for injection (*see* **Note 6**), into a small Petri dish in the Ficoll Injection Buffer. In the bottom of the Petri dish, a surface providing friction should be used to control the embryo positioning during injection (*see* **Note 7**). Arranging the embryos in rows will allow for a rapid and methodical approach to the injection procedure. Manipulating the embryos can be accomplished either with a fire-polished glass rod or dull forceps. The Petri dish containing the embryos may be placed on ice during the injection process. This offers a longer period of time for injection of a specific stage as the embryos develop more slowly. Also, the cooler temperature may increase the percentage survival of the injected embryos. An alternative to injecting embryos in Petri dishes, is to place embryos, in Ficoll Injection Buffer, directly on a piece of parafilm in groups of 10–20.

9. Bring the injection needle into close contact with the embryo and then, using a single fairly rapid movement, penetrate the surface membrane until the needle is just below the surface of the embryo (*see* **Fig. 4A**). Inject the selected volume of DNA solution into the embryo. Gently remove the needle and proceed with injection of the next embryo. It is useful to tilt the embryos so that the needle penetrates at a 90° angle to the surface of the embryo. This significantly reduces damage to the embryo and subsequent blebbing (*see* **Note 8**). Blebbing is the extrusion of cytoplasm and cellular material from the small hole in the fertilization membrane at the site of injection (**Fig. 4B**) (*see* **Note 9**). It is important not to tear the embryo at the site of injection (for choosing the site of injection, *see* **Note 10**) when removing the needle, as this can exacerbate blebbing. If leakage of cytoplasm or ripping of the fertilization membrane is observed regularly during the injection process, the tip of the micropipet may be the source of the problem and the needle may need to be replaced (*see* **Note 8**).

10. When an appropriate number of embryos has been injected—typically about 100 to 200 for each DNA construction—transfer them to a 13°C incubator. The embryos develop more slowly at 13°C and seem to survive the trauma of injection significantly better than embryos incubated at room temperature, although this effect is somewhat variable, probably depending on the batch of embryos and also on the injection needle. It is not unusual to observe death rates of 10–40%. About 2–12 h after injection, and before gastrulation, embryos should be transferred from the Ficoll Injection Buffer into incubation buffer (*see* **Note 11**). The embryos are now allowed to develop to the stage at which they will be assayed. Injected embryos can be incubated at any temperature from about 13°C to 22°C, but seem to survive better at lower temperatures (not below 10°C however).

11. After injection, the embryos must be monitored carefully. Dead or dying embryos seem to damage neighboring embryos and so it is important to remove necrotic embryos as soon as possible. These are readily identifiable by an uneven, marbled appearance of the pigmented hemisphere and later by the appearance of white necrotic tissue. Make sure to check the embryos early in the morning after overnight incubations because many of them may die overnight.

3.2. Transfection of DNA Constructions

1. Prepare DNA transfection mixture by combining plasmid DNA in water and transfection reagent at a ratio of 1:3 by mass (*see* **Note 12**). There is a range of different reagents that can be used and some of these are listed in **Note 13**). Concentration of plasmid DNA used should be relatively high, e.g., 4 µg/mL (*see* **Note 12**). We also include the FDA or RDA lineage tracer in the injection mixture (5 mg/mL) in order to detect the site of injection.
2. Load the injection needle with the transfection mixture in the same manner as described for standard DNA injections (see above) and use needles with the same properties.
3. Place embryos in 100% Steinberg's solution containing 25 µg/mL gentamycin. Anesthetize and immobilize the embryos with 3-aminobenzoic acid ethyl ester (methane sulfonate salt) at a concentration of 0.2 mg/mL. Immobilization is important to keep the embryo from moving and thereby expelling the injected transfection mixture.
4. Insert the micropipet tip into the desired embryonic tissue. In our case, we insert the micropipet under the epidermis at a 30° angle to the flank of the embryo and then inject the smallest volume possible (4.6 nL with the Nanoject system). At this point, a small bubble of injected material will raise the epidermis at the site of injection. Leaving the needle in place for a minute or so following the injection prevents leakage of the transfection mixture and allows a longer period of exposure of embryonic cells to the transfection mixture (*see* **Note 14**). When the needle is removed, the mixture can be observed to slowly leak out from the injection hole. Allow embryos to heal either at room temperature or at 13°C. The precise location of the transfections can be determined using fluorescence microscopy. Our experiments indicate that localized groups of 10–50 cells will subsequently express GFP after 24 h. The FDA or RDA used as lineage tracer appears to be taken up quite efficiently by the cells exposed to the transfection reagent and it helps mark the location of the transfection injection. Based on our preliminary experiments, it appears that approximately 1–10% of the cells that have taken up the rhodamine go on to express the GFP construction. **Figure 2** shows a comparison of the results of a standard DNA injection into a single blastomere of a 16-cell embryo (**Fig. 2A,B**) and transfection into the flank of a stage-23 tailbud embryo (**Fig. 2C,D**). Although this transfection protocol is not very efficient at present, it does allow expression of specific DNA constructions to be directed to extremely specific sites in the developing *Xenopus* embryo.

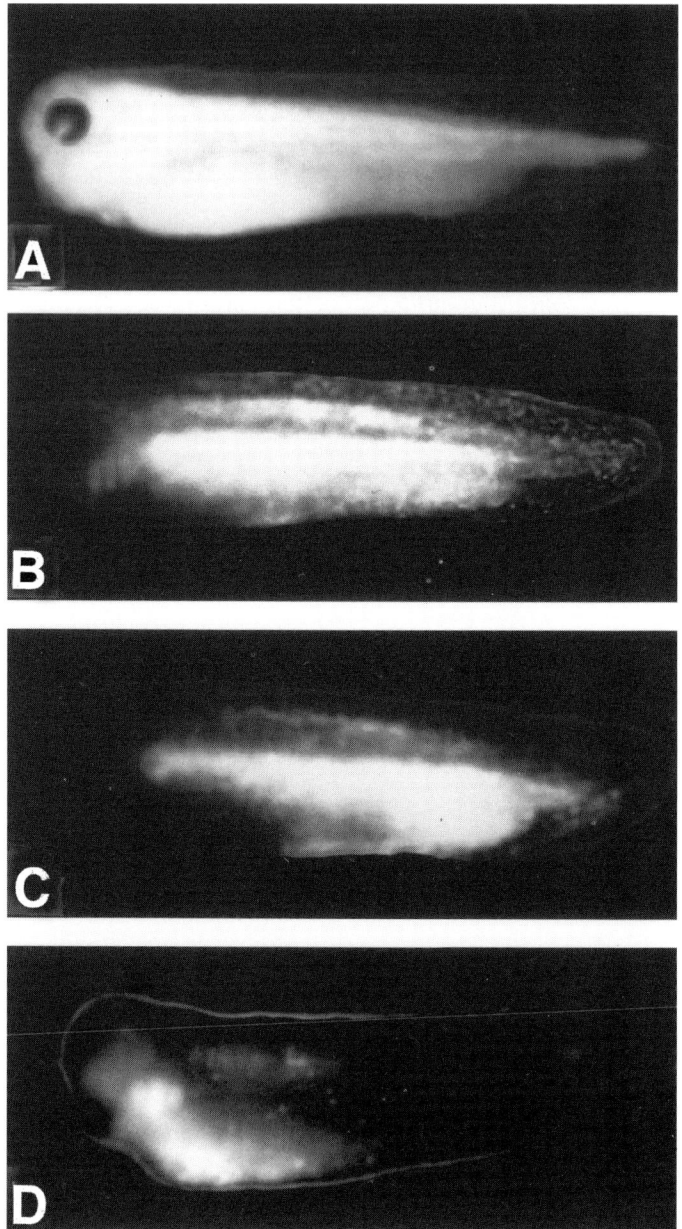

Fig. 3. Distribution of lineage tracer following injection into cleavage stage embryos. FDA was used as the tracer and all embryos were assayed by fluorescence at the tadpole stage (stage 36). **(A)** Injection at the 2-cell stage; **(B)** injection into the vegetal portion of a ventral blastomere at the 4-cell stage; **(C)** injection into the veg-

4. Notes

1. We have had good results with a Nanoject variable automatic injector (Drummond Scientific Co.) which is a very reliable and rather inexpensive injection apparatus. The injection volume of the Nanoject can be altered in 4.6 nL increments, starting from a low volume of 4.6 nL and extending up to 73.6 nL. This injector is small enough to be mounted on any simple micromanipulator (e.g., Narishige [Tokyo, Japan] MN-151 or Brinkman [Gravenzande, The Netherlands] manipulator or equivalent) allowing for the assembly of a very efficient but economical injection setup. Alternatives to the Nanoject include the pressurized gas microinjectors (available from companies like from Medical Systems, Narishige, and Nikon; *see* Chapter 10). These injectors are considerably more expensive and more complicated than the direct displacement Nanoject apparatus, but they have the advantage of allowing the injection of continuously variable volumes and also of much smaller volumes if necessary.

2. A number of different low-ionic-strength DNA buffers are probably suitable, but we have detected no advantage in using more complicated buffers. This solution usually also contains a lineage tracer. The inclusion of lineage tracers into the DNA injection solution serves to identify tissues that have received the injected plasmid DNA. Lineage tracers such as FDA and RDA have the advantage that they are relatively inert, can be visualized in living embryos, and can be fixed when conjugated with lysines. Tracer mRNAs are particularly useful for mRNA expression experiments because they control for RNase contamination and tend to diffuse at the same rate as the coinjected mRNA. One disadvantage of the β-galactosidase mRNA is that its expression is only detectable after fixation and is, therefore, not useful when living cells expressing mRNA must be followed. However, β-galactosidase can be directly visualized overlapping any color reaction following an *in situ* hybridization; this is not possible using fluorescent tracers.

   In addition, FDA or RDA in the DNA injection mixture provides color that can be used to visualize the DNA injection solution in the needle. This allows any leakage from the needle (excess positive pressure) or any backfilling (excess negative pressure) to become apparent. It is also useful when testing the apparatus. Periodically during a series of injections, a given volume of DNA injection solution should be injected directly into the Ficoll Injection Buffer to ensure that the Nanoject is functioning properly. The size of the injected bolus should remain constant throughout the injection series. Occasionally, the needle may become clogged and a test injection will reveal this.

   There is no consensus on the question of whether injected DNA should be linear or circular. There are some reports that expression is more efficient when

---

etal portion of a ventral blastomere at the eight-cell stage; (**D**) tracer is restricted to the heart, pharyngeal region, and liver (and a portion of the anterior somites) following injection into the CD2 blastomere at 16-cell stage.

linear templates are injected *(28,64)*. Whereas, supercoiled plasmids are known to persist until at least the late neurula stage *(25)*, linear DNAs have been shown to concatenate into high-molecular-weight complexes that integrate into the endogenous *Xenopus* genome, replicate, and may persist later during development *(16–19,24,65)*. Several laboratories have shown that circular plasmid DNA is amplified up to a hundred fold *(16–19,24)*; in contradiction, others have reported that circular plasmids are not appreciably amplified as free plasmid forms *(66,67)*. Although the results may differ for specific constructions, we have carried out a number of experiments with different promoters and different plasmid constructions and have not detected any consistent difference in the expression obtained from linearized and circular plasmid DNA.

3. The FDA and RDA lineage tracers and GFP protein produced from the injected mRNA may be viewed in the embryo under the appropriate wavelength of fluorescent light using a compound microscope. The β-gal protein produced from the injected mRNA is viewed using a standard light microscope following a color reaction (*see* **Note 4**).

4. β-Galactosidase color reaction; If injection of β-galactosidase mRNA is to be used for lineage tracing, embryos at the desired stage (up to stage 40) should be fixed using MEMPFA for 30 min. They should then be washed with PBS five times for 5 min each and placed in β-galactosidase staining solution (*see* **Subheading 2., item 15**). The color reaction can develop at 37°C for up to 3 h. Observe embryos every 30 min or so during this time and stop the color reaction when it is optimal. Then wash embryos with PBS three times for 5 min each, refix in MEMPFA for 30 min, and store in methanol at –20°C.

5. The amount of DNA injected is important for at least two reasons. First, increasing the amount of injected DNA increases the level of expression and decreases the degree of mosaicism (by allowing more cells to take up at least some DNA). This strategy works well for small amounts of injected DNA—up to about 250 pg—but becomes a problem with larger amounts. Above about 250 pg, the injected DNA exhibits a variable degree of amplification during subsequent embryonic development *(18,19)*. In some cases, the DNA can be amplified as much as 20–100-fold over injected levels. Although this amplification is not necessarily a problem, the variable degree of amplification between different individual embryos and different experiments can make quantitation of experimental results extremely difficult or impossible. For most purposes, therefore, it is useful to keep the amount of DNA injected below 200 pg, thereby reducing the likelihood of amplification artifacts.

The second concern is toxicity. In general, toxicity does not seem to be a problem, until the total amount of DNA injected reaches about 2 ng per injection. Although this effect does appear to be somewhat sequence specific, the presence of large amounts of plasmid DNA often causes the injected cells to undergo necrosis and die. If this occurs, cell death will become evident prior to gastrulation.

6. The developmental stage at which embryos should be injected varies between the 1-cell and the 64-cell stage (for details on fate mapping, see below). The stage chosen will generally reflect the desired spatial distribution. Earlier injection (for example, the one-cell stage) results in a much wider distribution of the injected

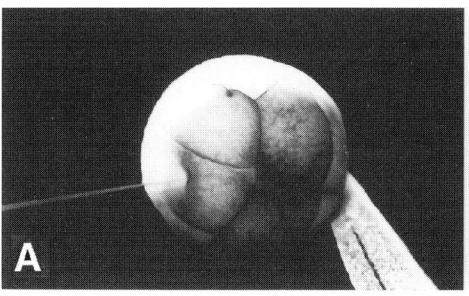 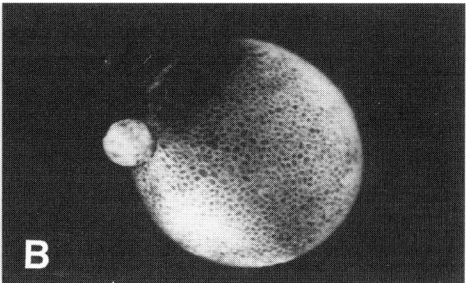

Fig. 4. Injection of cleavage stage *Xenopus* embryos. (**A**) Injection of 8-cell stage embryo into a dorsal animal blastomere. The needle has not yet penetrated the vitelline membrane. The forceps are used to hold the embryo in place. (**B**) A bleb observed at the site of injection of an embryo that has developed to stage 9.

material (*see* **Fig. 3A,B**), whereas later injection (for example, the 32-cell stage) results in a higher concentration of the injected material in a specific position in the embryo (*see* **Fig. 3C,D**). We have generally found that the best stage for tissue-directed injection is the 16-cell stage (*see* **Fig. 3D**). At this stage, the pigmentation and cleavage patterns of the blastomeres make it relatively easy to orient the embryo and direct the injection to desired tissues. In the majority of cases, darker pigmentation can be observed on ventral blastomeres and lighter pigmentation on dorsal blastomeres. Injection of DNA at the 8-cell stage, however, results in DNA expression in a much larger region of the embryo (as judged by the lineage tracer distribution) (*see* **Fig. 3C**). At the 32-cell stage, many embryos fail to cleave with perfect symmetry into four distinct tiers and it can be difficult to identify the desired blastomere for injection.

7. The membrane surrounding the embryo is rather tough and sometimes considerable pressure is required to penetrate the membrane and insert the needle into the underlying blastomere. Sometimes embryos tend to move in response to the pressure of the needle and it may be necessary to immobilize the embryo, either by holding gently with forceps or by supporting the embryo on a grid of Nitex mesh or, alternatively, by placing the embryo on paper with a rough surface such as 3MM (Whatmann, Fisher Scientific, Chicago, IL).

8. The correct shape of the tip of the needle is an important factor for smooth and successful injection. After the two-stage pull using the micropipet puller, the tip of the glass micropipet is broken manually with forceps at the point where the pipet becomes thin and flexible. Tips that are too blunt or irregularly barbed may cause excessive damage to the injected blastomere. We try to produce micropipets with a very small diameter (10–50 µm) and a slightly beveled tip. We find 25 µm to be optimal. The beveling allows for a smoother penetration of the fertilization membrane, without a serious indentation of the blastomere membrane upon application of pressure with the micropipet tip. The surface membrane should only slightly indent under the pressure of the needle during injection (*see* **Fig. 4A**).

9. Ficoll is often included in the injection buffer at a concentration of about 5% (w/v) to prevent leakage of the cytoplasm and subsequent blebbing of blastomeres by tightening the fertilization membrane around the embryo. Ficoll is an expensive reagent and therefore is only included in the buffer when necessary. We find that a Ficoll concentration of 3% is sufficient to reduce cytoplasmic leaking and blebbing. Quite apart from the embryonic damage caused by blebbing, the extruded cytoplasm can often contain a significant proportion of the injected DNA solution. This is particularly conspicuous when using a fluorescent lineage tracer.
10. Use fate mapping to direct expression to limited regions of the embryo. A complete fate map of the 32-cell *Xenopus* embryo has been established by Dale and Slack **(68)** and Moody **(69)**. Using the fate map information, ectopic expression of DNA constructs may be targeted to specific embryonic tissues by directed injection into specific blastomeres. Generally, we observe that targeting tissues using this fate map results in reliable distribution of injected material within the expected tissues. However, it must be noted that injected material is never restricted to a single tissue following injection of an individual blastomere at the 32-cell stage, as no single blastomere at the 32-cell stage contributes exclusively to any single tissue. For example, injections targeting the heart are carried out by injecting the C2 blastomere which is fated to contribute to the heart (*see* **ref. *68*** for blastomere nomenclature). Later in development, lineage tracer can be found in the heart region, but it is also found in the pharyngeal arches, the anterior lateral plate mesoderm, and in the notochord *(70)*. Injections targeting the kidney are carried out by injecting the C3 blastomere *(71)*. Injected material can later be found in the kidney and also in the somites and other derivatives of the lateral plate mesoderm. In addition, as mentioned earlier, we prefer to inject at the 16-cell stage rather than the 32-cell stage. However, blastomeres at this time contribute to a greater area of the developing embryo. To target the injected material more efficiently, injections are carried out in the portion of the 16-cell-stage blastomere that will eventually become the desired 32-cell-stage blastomere. For example, if targeting the heart when injecting at the 16-cell stage, the injection is carried out in the animal half of the CD2 blastomere, as this portion will eventually become part of the C2 blastomere.
11. Sometime prior to gastrulation, embryos should be removed from the Ficoll injection buffer and placed in Incubation Buffer. Embryos often do not gastrulate properly in a high-salt solution that contains Ficoll, and so it is important to remove them before this stage in development.
12. There are a number of parameters that affect the efficiency of the transfection technique. One factor is the ratio of DNA to transfection reagent. Generally, we use a ratio of 1:3 by mass (DNA:transfection reagent); for example, 0.5 µL of a 4 µg/µL DNA solution is mixed with 6 µL of a 1 µg/µL DOTAP solution (2 µg of DNA plus 6 µg of *N*-[1-(2,3-dioleoyloxy)propyl]-*N,N,N*-trimethyl ammonium methy sulfate [DOTAP]). This optimal ratio may vary for different constructions and a range of ratios should be assayed (1:2, 1:3, 1:4 DNA to transfection reagent).

13. A number of transfections reagents are commercially available. Examples include Tfx-10, Tfx-20, Tfx50, and Transfectam from Promega (Madison, WI), Lipofectamine and Lipofectin from Gibco BRL Life Technologies (Gaithersburg, MD), and DOTAP from Boehringer Mannheim/Roche (Indianapolis, IN). These reagents are generally more gentle than either the calcium phosphate or diethylaminoethyl (DEAE)–dextran methods and exhibit reduced cytotoxic effects. They are also advertised to show high rates of transfection for both DNA and RNA. We have successfully used both Transfectam and DOTAP.
14. A critical parameter for efficient transfection is to increase the time of exposure of the transfection mixture with the embryonic cells. The longer the treatment, the more likely the cells are to take up the plasmid DNA. These reagents are mostly designed to be used with cells in culture and suggested exposure times range from hours to days. Because these long exposures are not possible when a transfection mixture is injected into embryos, we try to extend each injection by leaving the needle inserted into the embryo for 1 min or so. This allows the transfection mixture to be in contact with the embryonic cells for as long as possible.

References

1. Kroll, K. L. and Amaya, E. (1996) Transgenic *Xenopus* embryos from sperm nuclear transplantations reveal FGF signaling requirements during gastrulation. *Development* **122,** 3173–3183.
2. Knox, B. E., Schlueter, C., Sanger, B. M., Green, C. B., and Besharse, J. C. (1998) Transgene expression in *Xenopus* rods. *FEBS Lett.* **423,** 117–121
3. Colman, A. (1975) Transcription of DNAs of known sequence after injection into eggs and oocytes of *Xenopus laevis*. *Eur. J. Biochem.* **57,** 85–96.
4. Mertz, J. E. and Gurdon, J. B. (1977) Purified DNAs are transcribed after microinjection into *Xenopus* oocytes. *Proc. Natl. Acad. Sci. USA* **74,** 1502–1506.
5. Rungger, D. and Turler, H. (1978) DNAs of simian virus 40 and polyoma direct the synthesis of viral tumor antigens and capsid proteins in *Xenopus* oocytes. *Proc. Natl. Acad. Sci. USA* **75,** 6073–6077.
6. Kressman, A., Clarkson, S. G., Telford, J. L., and Birnstiel, M. L. (1978) Transcription of *Xenopus* tRNA met and sea urchin histone DNA injected into *Xenopus* oocyte nucleus. *Cold Spring Harbor Symp. Quant. Biol.* **42,** 1077–1082.
7. Etkin, L. D. (1976) Regulation of lactate dehydrogenase (LDH) and alcohol dehydrogenase (ADH) synthesis in liver nuclei, following their transfer into oocytes. *Dev. Biol.* **52,** 201–209.
8. Probst, E., Kressmann, A., and Birnstiel, M. L. (1979) Expression of sea urchin histone genes in the oocyte of *Xenopus laevis*. *J. Mol. Biol.* **135,** 709–732.
9. De Robertis, E. M. and Mertz, J. (1977) Coupled transcription-translation of DNA injected into *Xenopus* oocytes. *Cell* **12,** 175–182.
10. Bakken, A., Morgan, G., Sollner-Webb, B., Roon, J., Busby, S., and Reeder, R. (1982) Mapping of transcription initiation and termination signals in *X. laevis* ribosomal DNA. *Proc. Natl. Acad. Sci. USA* **79,** 56–60.

11. Trendelenburg, M. F. and Gurdon, J. B. (1978) Transcription of cloned *Xenopus* ribosomal genes visualized after injection into oocyte nuclei. *Nature* **276,** 292–294.
12. Trendelenburg, M. H., Zentgraf, H., Franke, W. W., and Gurdon, J. B. (1978) Transcription patterns of amplified Dytiscus genes coding for ribosomal RNA after injection into *Xenopus* oocyte nuclei. *Proc. Natl. Acad. Sci. USA* **75,** 3791–3795.
13. Brown, D. D. and Gurdon, J. B. (1977) High fidelty transcription of 5S DNA injected into *Xenopus* oocytes. *Proc. Natl. Acad. Sci. USA* **74,** 2064–2068.
14. McKnight, S. L., Gavis, E. R., Kingsbury, R., and Axel, R. (1981) Analysis of transcriptional regulatory signals of the HSV thymidine kinase gene: identification of an upstream control region. *Cell* **25,** 385–398.
15. Etkin, L. D. and Maxson, R. E., Jr. (1980) The synthesis of authentic sea urchin transcriptional and translational products by sea urchin histone genes injected into *Xenopus laevis*. *Dev. Biol.* **75,** 13–25.
16. Bendig, M. M. (1981) Persistence and expression of histone genes injected into *Xenopus laevis* eggs in early development. *Nature* **292,** 65–67.
17. Etkin, L. D. (1982) Analysis of the mechanisms involved in gene regulation and cell differentiation by microinjection of purified genes and somatic cell nuclei into amphibian oocytes and eggs. *Differentiation* **21,** 149–159.
18. Rusconi, S. and Schaffner, W. (1981) Transformation of frog embryos with a rabbit β-globin gene. *Proc. Natl. Acad. Sci. USA* **78,** 5051–5055.
19. Etkin, L. D., Pearman, B., Roberts, M., and Bektesh, S. L. (1984) Replication, integration and expression of exogenous DNA injected into fertilized eggs of *Xenopus laevis*. *Differentiation* **26,** 194–202. .
20. Newport, J. W. and Kirschner, M. W. (1982) A major developmental transition in early *Xenopus* embyros: I. Characterization and timing of cellular chages at the midblastula stage. *Cell* **30,** 675–686.
21. Newport, J. W. and Kirschner, M. W. (1982) A major developmental transition in early *Xenopus* embyros: II. Control of the onset of transcription. *Cell* **30,** 687–696.
22. Forbes, D. J., Kirschner, M. W., and Newport, J. W. (1983) Spontaneous formation of nucleus-like structures around bacteriophage DNA microinjected into *Xenopus* eggs. *Cell* **34,** 13–23.
23. Shiokawa, K., Sameshima, M., Tashiro, K., Miura, T., Nakakura, N., and Yamana, K. (1986) Formation of nucleus-like structures in the cytoplasm of lambda injected fertilized eggs and its partitioning into blastomeres during early embryogenesis of *Xenopus laevis*. *Dev. Biol.* **116,** 539–542.
24. Etkin, L. D. and Pearman, B. (1987) Distribution, expression, and germ line transmission of exogenous DNA sequences following microinjection in *Xenopus laevis* eggs. *Development* **99,** 15–23.
25. Krieg, P. A. and Melton, D. A. (1985) *Development*al regulation of a gastrula-specific gene injected into fertilized *Xenopus* eggs. *EMBO J.* **4,** 3463–3471.
26. Krieg, P. A. and Melton, D. A. (1987) An enhancer responsible for activating transcription at the midblastula transition in *Xenopus* development. *Proc. Natl. Acad. Sci. USA* **84,** 2331–2335.

27. Mohun, T. J., Garret, N., and Gurdon, J. B. (1986) Upstream sequences required for tissue-specific activation of the cardiac actin gene in *Xenopus laevis* embyros. *EMBO J.* **5,** 3185–3193.
28. Wilson, C., Cross, G. S., and Woodland, H. R. (1986) Tissue-specific expression of actin genes injected into *Xenopus* embryos. *Cell* **47,** 589–599.
29. Jonas, E. A., Snape, A. M., and Sargent, T. D. (1989) Transcriptional regulation of a *Xenopus* embryonic epidermal keratin gene. *Development* **106,** 399–405.
30. Krone, P. H. and Heikkila, J. J. (1989) Expression of microinjected HSP70/CAT and HSP30/CAT chimeric genes in developing *Xenopus laevis* embryos. *Development* **106,** 271–281.
31. Brennan, S. M. (1990) Transcription of endogenous and injected cytoskeletal actin genes during early embryonic development in *Xenopus laevis*. *Differentiation* **44,** 111–121.
32. Gong, S. G., Reddy, B. A., and Etkin, L. D. (1995) Two forms of *Xenopus* nuclear factor 7 have overlapping spatial but different temporal patterns of expression during development. *Mech. Dev.* **52,** 305–318.
33. Nakamura, H., Tashiro, K., and Shiokawa, H. (1996) Isolation of *Xenopus* HGF gene promoter and its functional analysis in embryos and animal caps. Roux Arch *Dev. Biol.* **205,** 300–310.
34. Mohun, T. J., Brennan, S., Dathan, N., Fairman, S., and Gurdon, J. B. (1984) Cell type-specific activation of actin genes in the early amphibian embryo. *Nature* **311,** 716–721.
35. Liebham, D., Wong, M. W., Cheng, T. C., Schroeder, S., Weil, P. A., Olson, E. N., et al. (1994) Binding of TFIID and MEF2 to the TATA element activates transcription of the *Xenopus* MyoDa promoter. *Mol. Cell. Biol.* **14,** 686–699.
36. Steinbeisser, H., Alonso, A., Epperlein, H. H., and Trendelenburg, M. F. (1989) Expression of mouse histone H1^0 promoter sequences following microinjection into *Xenopus* oocytes and developing embryos. *Int. J. Dev. Biol.* **33,** 361–368.
37. Mayor, R., Essex., L. J., Bennett, M. F., and Sargent, M. G. (1993). Distinct elements of the *xsna* promoter are required for mesodermal and ectodermal expression. *Development* **119,** 661–671.
38. Chan, A. P. and Gurdon, J. B. (1996). Nuclear transplantation from stably transfected cultured cells of *Xenopus*. *Int. J. Dev. Biol.* **40,** 441–451.
39. Gao, X., Kuiken, G. A., Baarends, W. M., Koster, J. G., and Destree, O. H. (1994) Characterization of a functional promoter for the *Xenopus* wnt-1 gene on vivo. *Oncogene* **9,** 573–581.
40. Batni, S., Scalzetti, L., Moody, S. A., and Knox, B. E. (1996).Characterization of the *Xenopus* rhodopsin gene. J. Biol. Chem. **271,** 3179–3186.
41. Weber, H., Holewa, B., Jones, E. A., and Ryffel, G. U. (1996) Mesoderm and endoderm differentiation in animal cap explants: identification of the HNF4-binding site as an activin A responsive element in the *Xenopus* HNF1alpha promoter. *Development* **122,** 1975–1984.
42. Croissant, J. D., Kim, J. H., Eichele, G., Goering, L., Lough, J., Prywes, R., et al. (1996) Avian serum response factor expression restricted primarily to

muscle cell lineages is required for alpha-actin gene transcription. *Dev. Biol.* **177,** 250–264.
43. Hedgepeth, C. M., Conrad, L. J., Zhang, J., Huang, H. C., Lee, V. M., and Klein, P. S. (1997) Activation of the Wnt signaling pathway: a molecular mechanism for lithium action. *Dev. Biol.* **185,** 82–91.
44. Kaufmann, E., Paul, H., Friedle, H., Metz, A., Scheucher, M., Clement, J. H., et al. (1996) Antagonistic actions of activin A and BMP-2/4 control dorsal lip-specific activation of the early response gene XFD-1' in *Xenopus laevis* embryos. *EMBO J.* **15,** 6739–6749.
45. Johnson, A. D., and Krieg, P. A. (1994) pXeX, a vector for efficient expression of cloned sequences in *Xenopus* embryos. *Gene* **147,** 223–226.
46. Vize, P. (1996) DNA sequences mediating the transcription response of the Mix.2 homeobox gene to mesoderm induction. *Dev. Biol.* **177,** 226–231.
47. Weber, H., Strandmann, E.P., Holewa, B., Bartkowski, S., Zapp, D., Zoidl, C., et al. (1996) Regulation and function of the tissue-specific transcription factor HNF1 alpha (LFB1) during *Xenopus* development. *Int. J. Dev. Biol.* **40,** 297–304.
48. Christian, J. L. and Moon, R. T. (1993) Interactions between *Xwnt*-8 and Spemann organizer signaling pathways generate dorsoventral pattern in the embryonic mesoderm of *Xenopus*. *Genes Dev.* **7,** 13–28.
49. Jones, C. M., Dale, L., Hogan, B. L., Wright, C. V., and Smith, J. C. (1996) Bone morphogenetic protein-4 (BMP-4) acts during gastrula stages to cause ventralization of *Xenopus* embryos. *Development* **122,** 1545–1554.
50. Cleaver, O., Tonissen, K. F., Saha, M. S., and Krieg, P. A. (1997) Neovascularization of the *Xenopus* embryo. *Dev. Dynam.* **210,** 66–77.
51. Turner, D. L. and Weintraub, H. (1994) Expression of achaete-scute homolog 3 in *Xenopus* embryos converts ectodermal cells to a neural fate. *Genes Dev.* **8,** 1434–1447.
52. Kuhl, M., Finnemann, S., Binder, O., and Wedlich, D. (1996) Dominant negative expression of a cytoplasmically deleted mutant of XB/U-cadherin disturbs mesoderm migration during gastrulation in *Xenopus laevis*. *Mech. Dev.* **54,** 71–82.
53. Wallingford, J. B., Carroll, T. C., and Vize, P. D. (1998) Precocious expression of the Wilms' Tumor *Gene xWT1* inhibits embryonic kidney development in *Xenopus laevis*. *Dev. Biol.* **202,** 103–112.
54. Fu, Y., Wang, Y., and Evans, S. M. (1998) Viral sequences enable efficient and tissue-specific expression of transgenes in *Xenopus*. *Nature Biotech.* **16,** 253–257.
55. Sanes, J. R., Rubenstein, J. L., and Nicolas, J. F. (1986) Use of a recombinant retrovirus to study post-implantation cell lineage in mouse embryos. *EMBO J.* **5,** 3133–3142.
56. Price, J. (1987) Retroviruses and the study of cell lineage. *Development* **101,** 409–419.
57. Luskin, M. B., Pearlman, A. L., and Sanes, J. R. (1988) Cell lineage in the cerebral cortex of the mouse studied in vivo and in vitro with a recombinant retrovirus. *Neuron* **1,** 635–647.
58. Holt, C. E., Garlick, N., and Cornel, E. (1990) Lipofection of cDNAs in the embryonic vertebrate central nervous system. *Neuron* **4,** 203–214.

59. Moon, R.T. and Christian, J. L. (1989) Microinjection and expression of synthetic mRNAs in *Xenopus* embryos. *Technique* **1,** 76–78.
60. Kay, B.K. (1991) Injection of oocytes and embryos, in *Methods in Cell Biology:* Xenopus laevis—*Practical Uses in Cell and Molecular Biology* **36,** 663–669.
61. Chalfie, M., Tu, Y., Euskirchen, G., Ward, W. W., and Prasher, D. C. (1994) Green fluorescent protein as a marker for gene expression. *Science* **263,** 802–805.
62. Newmeyer, D. D., and Wilson, K. L. (1991) Egg extracts for nuclear import and nuclear assembly reactions, in *Methods in Cell Biology:* Xenopus laevis—*Practical Uses in Cell and Molecular Biology,* **36,** 607–634.
63. Vize, P. D., Melton, D. A., Hemmati-Brivanlou, A., and Harland, R. M. (1991) Assays for gene function in developing *Xenopus* embryos, in *Methods in Cell Biology:* Xenopus laevis—*Practical Uses in Cell and Molecular Biology* **36,** 367–387.
64. Snape, A. M., Jonas, E. A., and Sargent, T. D. (1990) KTF-1, a transcriptional activator of *Xenopus* embryonic keratin expression. *Development* **109,** 157–165.
65. Marini, N. J., Etkin, L. D., and Benbow, R. M. (1988) Persistence and replication of plasmid DNA microinjected into early embryos of *Xenopus laevis*. *Dev. Biol.* **127,** 421–434.
66. Bendig, M. M. and Williams, J. G. (1981) Replication and expression of *Xenopus laevis* globin genes injected into fertilized *Xenopus* eggs. *Nature* **292,** 65–67.
67. Andres, A., Muellener, D. B., and Ryffel, G. U. (1984) Persistence, methylation and expression of vitellogenin gene derivatives after injection into fertilized eggs of *Xenopus laevis*. *Nucleic Acids Res.* **12,** 2283–2302.
68. Dale, L. and Slack, J. M. W. (1987) Fate map for the 32-cell stage of *Xenopus laevis*. *Development* **99,** 527–551.
69. Moody, S. A. (1987) Fates of the blastomeres of the 32-cell-stage *Xenopus* embryo. *Dev. Biol.* **122,** 300–319.
70. Cleaver, O. B., Patterson, K. D., and Krieg, P. A. (1996) Overexpression of the *tinman*-related genes *XNkx-2.5* and *XNkx-2.3* in *Xenopus* embryos results in myocardial hyperplasia. *Development* **122,** 3549–3556.
71. Wallingford, J. B., Seufert, D. W., Virta, V. C., and Vize, P. D. (1997) p53 activity is essential for normal development in *Xenopus*. *Curr. Biol.* **7,** 747–757.

13

Promoter Analysis in Zebrafish Embryos

Jos Joore

1. Introduction

Analysis of promoter regulation is a powerful strategy for the study of regulatory interactions between genes that play a role in embryonic development. It allows investigation of the molecular mechanisms that underlay temporal and spatial expression of a specific gene. In addition, one can study combinatorial signaling processes that are involved in the regulation of developmentally important genes. In general, promoter analysis involves the identification of specific regulatory DNA sequences, promoter, and enhancer elements, which are important for transcriptional activity of a gene. Enhancer elements are *cis*-acting elements that modulate gene expression in a time- and tissue-specific fashion *(1)*. The first phase in the analysis of a promoter involves the cloning of promoter fragments of different sizes into constructs that contain a reporter gene that allows evaluation of promoter activity (**Fig. 1**). Through analysis of promoter activity of such a deletion series, specific enhancer elements, which modulate promoter activity, can be identified. Subsequently, regulatory sequences can be dissected further by cloning smaller fragments into a reporter vector containing a heterologous, constitutively active promoter, for example the herpes simplex virus *thymidine kinase* promoter. At the highest resolution, single basepair substitutions in the context of the wild-type promoter may reveal the requirement of specific elements for regulation of the gene under study (**Fig. 1**).

The choice of methods for in vivo promoter analysis in embryos is determined by the specific characteristics of the model system. In many experimental animals, such as the mouse, *Drosophila, C. elegans*, and, recently, *Xenopus* embryos, transgenic animals can be routinely generated *(2–5)*. In transgenic animals, the promoter construct is stably integrated into the genome, resulting

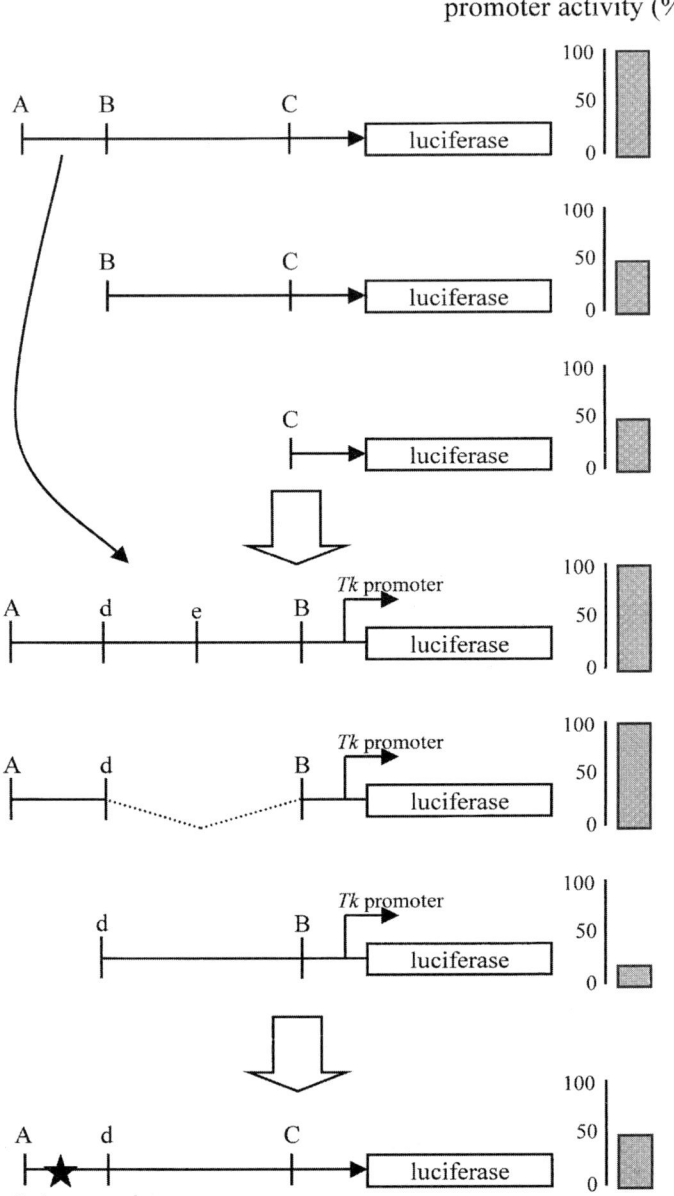

Fig. 1. Typical example of a strategy to analyze a promoter. A promoter deletion series is cloned into a luciferase reporter vector and analyzed for activity (top three constructs), which is indicated in the bar graphs This identifies a region A–B that is important for promoter activity. This region and various subfragments of this region are cloned into a reporter vector driven by the herpes simplex virus *thymidine kinase*

in all cells of the embryo containing the same number of copies. By contrast, in transient experiments, the distribution of the construct within the embryo is often mosaic, with some cells containing many copies and some cells containing no construct at all. This results in mosaic expression patterns, which are sometimes difficult to analyze. An additional complication is ectopic promoter activity, which is caused by high copy numbers of construct in the cell. Mosaiscism often leads to a high variation between individual embryos, rendering analysis difficult. Sometimes, mosaiscism can be useful (e.g., in *Xenopus*, where injection of single blastomeres at the 32-cell-stage targets the construct into specific regions of the embryo) *(6)*.

Although transgenesis is feasible in zebrafish, it is time-consuming and requires special facilities *(7)*. Thus, for the analysis of a large series of promoter constructs in zebrafish, the transient introduction of constructs by microinjection is preferable, even though this inevitably results in mosaiscism. Part of the variation caused by the mosaic distribution of constructs can be overcome by coinjecting a construct that serves as an internal control, assuming that the control construct displays the same degree of mosaiscism. Alternatively, if the promoter under analysis is regulated by soluble factors and is active during early developmental stages, preferably before gastrulation, a different approach is feasible in zebrafish. According to this method, constructs are introduced into embryos by microinjection, followed by the dissociation of a number of injected embryos, yielding a culture of mixed cells, many of which contain the injected construct *(8)*. These cultures are incubated with soluble factors, and subsequently promoter activity is determined. An important advantage of this method is that the variation between injected embryos does not influence the reproducibility of the results. Furthermore, it allows the use of pharmacological inhibitors and activators, which is often impossible in intact embryos because of permeability problems (unpublished observations). Drawbacks of this method are that the promoter is analyzed outside the normal embryonic context and that its application is restricted to a subset of scientific questions specifically involving promoters that are transcriptionally active during early developmental stages and that are regulated by soluble factors.

In this chapter, I will describe both methods for promoter analysis in zebrafish embryos. First, DNA microinjection is described briefly, followed by the description of quantitative analysis of promoter activity in intact

---

(*Tk*) promoter, which identifies the fragment Ad as the most important region for transcriptional activity. Finally, a point mutation in the full-length promoter shows that a specific enhancer element in Ad plays an important role in regulation of the transcriptional activity of the promoter (bottom construct).

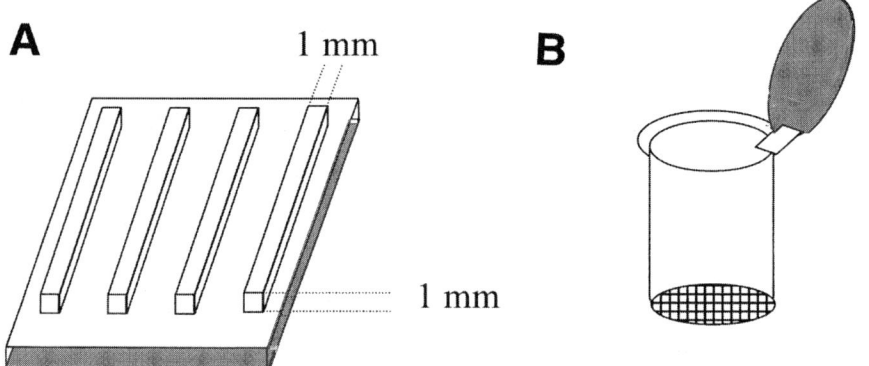

Fig. 2. **(A)** A plastic mold used to make troughs in agarose for mounting embryos during microinjection. (Modified from **ref. 11**.) **(B)** A chorion sieve made from the top half of an Eppendorf tube with nylon mesh attached to the bottom.

embryos and dissociated embryonic cells using firefly luciferase as a reporter gene. Qualitative spatial analysis in intact embryos is described using wholemount lacZ (β-galactosidase) staining.

2. Materials

2.1. Microinjection

1. Glass capillaries, 1-mm outer diameter (e.g., GC100TF, Clark Electromedical Instruments, Reading, UK).
2. A pipet puller, such as the Sutter P-80/PC (Novato, CA).
3. Injection buffer: 10 mM HEPES pH 7.0, 50 mM KCl, 0.1 mM EDTA pH 8.0.
4. Embryo medium (standard culture medium for embryos without chorion as described by Westerfield [9]).
5. Plastic mold for making five 0.1 × 0.1 × 3-cm slots in agarose in a Petri dish (**Fig. 2**).
6. 1.5% Agarose in embryo medium.
7. An injection binocular stereomiscroscope equipped with a micromanipulator and a pressure injector. The binocular should have a magnification range between 10× and 50×. The micromanipulator should be stable, although state-of-the-art equipment with very fine controls is not strictly required. The pressure injector should have a fine pressure control in the range of 0–100 psi and a control for the injection time. A "clear" function that delivers a short high-pressure pulse can be useful to clear a clogged pipet. The pipet holder should be suitable for 1-mm capillaries.

2.2. Quantitative Promoter Analysis in Dissociated Blastula Cells

1. Phosphate buffered saline (PBS, without calcium and magnesium) is obtained as tablets from Sigma (Zwijndrecht, Holland) (#1000-3) dissolved in water and autoclaved.

2. A chorion sieve (**Fig. 2**) is made by cutting off the bottom half of an Eppendorf tube and melting a piece of fine nylon mesh to the tube by using a hot surface (e.g., a flatiron).
3. Leibovitz L15 medium is obtained from Sigma (#L-4386) and stored at –20°C or for up to 1 mo at 4°C.
4. Lysis buffer: 1% Triton X-100 in 25 mM glycylglycine, 15 mM magnesium sulfate, 4 mM ethylene glycol-bis(β-aminoethyl ether)-$N,N,N'N'$-tetraacetic acid (EGTA). Adjust the pH to 7.6 and store at 4°C for up to 2 mo. Immediately before use, 1 mM dithiothreitol (DTT) is added.
5. For luciferase activity determination, any method available may be used (*see* **Note 1**). I have obtained good results using the method described by Brasier et al. *(10)* and a Lumac M2010A luminometer (Landgraaf, Holland), as well as using a TopCount scintillation counter with luciferase assay reagent supplied by Packard (Meriden, CT).

2.3. Qualitative and Quantitative Promoter Analysis in Intact Embryos

1. An Eppendorf micropestle for pottering samples in an Eppendorf tube.
2. LacZ assay buffer: 100 mM potassium phosphate buffer pH 7.8, 10 mM KCl, 1 mM MgSO$_4$ is stored at room temperature. Immediately before use, 3.8 µL/mL β-mercaptoethanol is added.
3. LacZ assay mix is prepared immediately before use by mixing for each sample 100 µL 100 mM potassium phosphate buffer pH 7.8, 24 µL lacZ assay buffer, 66 µL distilled water, and 0.16 mg *o*-nitrophenyl β-D-galactopyranoside (ONPG, Sigma #N-1127, store at –20°C and protect from light).
4. Fixative is freshly made before use by mixing 694 µL formaldehyde (36%), 320 µL glutaraldehyde, and 2 µL NP-40 in a final volume of 10 mL in PBS. Keep on ice until use.
5. X-Gal Solution: 4% (w/v) 5-bromo-4-chloro-indolyl β-D-galactopyranoside (Sigma #B-4252) in dimethylformamide is stored at –20°C, protected from light.
6. LacZ staining buffer is made freshly before use by adding 20 µL 200 mM K$_4$Fe(CN)$_6$, 20 µL 200 mM K$_3$Fe(CN)$_6$, 2 µL 1 M MgCl$_2$, 20 µL 4% X-Gal solution, and 938 µL phosphate-buffered saline (PBS) for every two samples. 200 mM K$_4$Fe(CN)$_6$ should be prepared freshly before use; 200 mM K$_3$Fe(CN)$_6$ is stored at –20°C.
7. 70% glycerol in PBS.
8. Murray's clearing solution is made up by mixing one volume of benzoyl alcohol and two volumes of benzoyl benzoate. *Caution*: toxic. Store in a dark bottle at room temperature.

3. Methods
3.1. Microinjection

1. Pull microinjection pipets on a suitable apparatus, using 1-mm borosilicate glass capillaries with filament (e.g., GC100TF from Clark Electromedical Instru-

ments). Be sure to adjust the pipet puller to obtain pipets with a taper long enough to penetrate the chorion and the embryo. Avoid making the taper too long, as this hinders penetration of the chorion because of flexibility of the needle.
2. Break the tip of the pipet, this can be achieved in two ways. Most reproducible results are obtained using a watchmaker's forceps under a microscope equipped with a micrometer. Generally, 2.5 to 5-µm tips give good results. Alternatively, pipets can be broken under the injection binocular by breaking off a fixed length of the tip of the pipet using a watchmaker forceps.
3. Mount the embryos in troughs made with a plastic mold in agarose in a Petri dish, covered with embryo medium *(11)*. The plastic mold is designed to make five 1 mm wide, 1 mm deep troughs in an 85-mm Petri dish filled with 25 mL of 1.5% agarose in embryo medium. Let the mold float on top of the liquid, taking care not to trap air bubbles. Remove the plastic mold after the agarose has set and cover the agarose with a thin layer of embryo medium. Allow the dish to cool to room temperature before injection.
4. Select a number of eggs that have just started cleaving (*see* **Note 2**) and mount the eggs in an agarose trough with the blastomeres facing upward by gentle pushing with a blunt forceps. The embryos in their chorions should fit tightly in the troughs.
5. Dilute the DNA in injection buffer at a concentration of 1–20 µg/mL (*see* **Notes 3 and 4**). The optimal concentration should be determined experimentally. Load the pipet using an Eppendorf microloader or an ordinary yellow tip. Gently tap the pipet to let the solution flow into the tip and mount the pipet onto the pipet holder.
6. Check the injection volume using a binocular microscope equipped with a micrometer. To determine the injection volume, inject into an embryo or into a drop of mineral oil and measure the size of the injected droplet in microns. Assuming that the injected droplet is spherical, the volume in picoliters is calculated as $(size)^3/1910$.
7. To inject, force the pipet through the chorion, using the controls of the micromanipulator, and enter the embryo at the border between the blastomeres and the yolk cell. The embryo will turn slightly, positioning the pipet into one of the blastomeres. Inject approximately 250–500 pL into a blastomere and gently withdraw the pipet from the embryo. Alternatively, for experiments using dissociated blastula cells, the DNA may be injected at the border between the blastomeres and the yolk cell, which is generally faster than blastomere injection.

3.2. Promoter Analysis in Dissociated Blastula Cells

This method has been successfully applied to analyze the regulation of the zebrafish *goosecoid* promoter by activin and basic Fibroblast Growth Factor *(8)*.

1. Incubate 40–80 embryos injected with a promoter-luciferase construct at 28.5–31°C until the sphere stage (4 h postfertilization [h.p.f.]).

2. Transfer the embryos into an Eppendorf tube in 500 µL PBS.
3. Flick the tube to disperse the embryos and gently pull the embryos through a 20-gauge injection needle using a 1-mL syringe. This will break the chorions, triturate the embryos and dissociate the cells. Gently empty the syringe into the Eppendorf tube. Repeat the trituration three to five times until no intact chorions are observed in the suspension. Avoid using excessive force, as this will result in a loss of cells due to shearing (see **Note 5**).
4. When all embryos are dissociated, remove chorion debris by emptying the syringe through the chorion sieve into a clean Eppendorf tube. Rinse the syringe and the mesh with another 500 µL PBS.
5. Pellet the cells by spinning for 4 min at 400g in a microfuge and remove the supernatant. To avoid disturbing the pellet, leave approximately 10 µL of the supernatant in the tube.
6. Wash the cells with 1 mL Leibovitz L15 medium at room temperature. Finally, take up the cells in 10 µL Leibovitz L15 medium for every 2.5 embryo. When checked under a microscope, the suspension should contain single cells, some small aggregates of two to three cells and yolk platelets, which are recognized by the absence of a nucleus.
7. Transfer 10-µL aliquots of the cell suspension into clean Eppendorf tubes. It is absolutely crucial to keep the cells in suspension by gently flicking the tube before removal of each aliquot. To avoid systematic errors, the tubes should be filled in a randomized order. Assays should be set up in triplicate.
8. Add factors diluted in Leibovitz L15 medium and adjust the volume to 20 µL.
9. Mix the contents by gentle tapping and incubate the cells at 26°C for 2 h. An incubation time of two hours was sufficient to assay the induction of the zebrafish *goosecoid* promoter by activin and other factors *(8)*. Depending on the experimental setup, the incubation time may be varied. For some pharmacological inhibitors, such as the PKA inhibitor H89 or the protein synthesis inhibitor cycloheximide, a preincubation is required for maximum activity. In this case, the inhibitor is added 10–30 min before final addition of activating factors.
10. Cells are lysed to release luciferase protein by addition of 150 µL ice-cold lysis buffer and shaking for 5 min at room temperature in an Eppendorf shaker. Store the samples on ice. For optimal reproducibility, process the samples immediately for determination of luciferase activity. Luciferase assays are performed on 10–75 µL of lysate using methods available (see **Note 1**).

3.3. Promoter Analysis in Intact Embryos: Quantitative Analysis Using Luciferase as a Reporter Gene

To analyze promoter activity in intact embryos, several precautions should be taken to minimize the variability between samples. First, injections should be carried out with great precision. An ocular micrometer is used to monitor the injection volume, which should be kept as constant as possible. Avoid injecting volumes smaller than 250 pL, as these are hard to control accurately. Discard embryos in which the injected solution leaks out of the blastomere into the yolk cell. Second, for each construct a minimum of 3 samples of 10 embryos each should be analyzed in one experiment. If multiple constructs or combinations of constructs are used, these should preferably be assayed simultaneously. Third, an internal control can be included to normalize the data. To this end, 5 µg/mL of a lacZ expression construct driven by a strong, constitutive promoter is added to the injected DNA solution.

1. Inject embryos as described above (**Subheading 3.1.**) and allow them to develop until the appropriate developmental stage.
2. Remove embryos that show abnormal development, as these often show extremely high promoter activity, which may obscure the results.
3. Transfer samples of 10 randomly picked embryos into Eppendorf tubes and remove the embryo medium using a syringe equipped with a 26-gauge needle.
4. Lyse the embryos in 200 µL of ice-cold lysis buffer using an Eppendorf micropestle. Be sure to use tubes in which the micropestle fits perfectly, otherwise incomplete lysis may occur. Rinse the micropestle between samples to avoid cross-contamination.
5. Vortex the samples, put the tubes on ice and leave for 10 min.
6. Spin down debris at maximum speed in a microfuge at 4°C. Store the samples on ice and proceed with β-galactosidase and luciferase assays immediately.
7. For each sample, in a well of a 96-well plate, mix 190 µL lacZ assay mix and 50 µL embryo lysate.
8. Incubate at 37°C until a light yellow color is observed. Depending on the developmental stage of the embryos and the promoter used in the lacZ expression construct, this may take 30 min to several hours.
9. Determine the optical density at 420 nm in a microplate reader.
10. Determine luciferase activity in 10–75 µL lysate using methods available (*see* **Note 1**).
11. Calculate the mean of the optical densities as obtained from the LacZ assays. The corrected luciferase value for a specific sample is calculated as follows: [corrected sample luciferase activity] = [sample luciferase activity] × [mean LacZ optical density]/[sample LacZ optical density] (*see* **Note 6**).

3.4. Promoter Analysis in Intact Embryos: Spatial Analysis Using β-Galactosidase as a Reporter Gene

For spatial analysis of promoter activity, promoter fragments should be cloned into a vector containing the β-galactosidase (lacZ) gene (*see* **Note 7**). In these experiments, it is important to carefully optimize the amount of injected construct first. Excess construct will often result in ectopic expression, which obscures the pattern of activity, whereas too little injected DNA will result in low, highly mosaic patterns. Injections of 5 pg of DNA will generally serve as a good starting point. It is important to note that spatial analysis of promoter activity should, in fact, precede quantitative analysis whenever possible, to prevent misinterpretations resulting from ectopic promoter activity in luciferase experiments.

1. Inject embryos (*see* **Subheading 3.1.**) with a promoter construct driving the bacterial β-galactosidase (lacZ) gene and allow to develop to the appropriate developmental stage.
2. Transfer the embryos to a small Petri dish, with exception of those that show abnormal development.
3. Remove the embryo medium and fix the embryos in 1–5 mL fixative for 30 min at 4°C.
4. Following fixation, rinse the embryos twice in PBS for 5 min and remove the chorions using two sharp watchmaker's forceps (*see* **Note 8**).
5. Fill an Eppendorf tube with 500 µL lacZ staining solution for every sample and transfer the embryos in a small volume of PBS into the staining solution. This is best achieved by swirling the Petri dish to collect the embryos in the centre and sucking them into a cut-off wide-bore Pasteur pipet. Allow the embryos to settle to the bottom of the pipet, submerge the tip of the pipet into the staining solution without trapping air bubbles and gently push out the embryos. This prevents the embryos from floating on the surface of the staining solution.
6. Stain overnight at 30°C in the dark.
7. Transfer the embryos in a small volume to a small Petri dish with PBS and rinse once with PBS.
8. The stained embryos are examined and photographed against a white background with epi-illumination. To facilitate orientation, the embryos may be transferred to 70% glycerol in PBS. Alternatively, embryos may be transferred to methanol through a graded series (25%, 50%, 75%, and twice 100% in PBS) and subsequently to Murray's clearing solution (*see* **Note 9**). In embryos cleared in this way, detection of weak lacZ staining is improved (*see* **Note 10**).
9. Depending on the promoter, results may be analyzed in various ways. If ectopic activity and mosaicism are low, a careful comparison of stained embryos will reveal the general pattern of activity (*see* **Note 11**). Embryos that show a pattern that reflects the overall pattern are used for presentation of the results. Alternatively, a number of embryos are photographed in exactly the same orientation and all stained cells are indicated in a schematic drawing. In this way, the patterns of several embryos are combined to produce a reliable representation of the overall activity of the promoter, even if activity is low and highly mosaic.

4. Notes

1. Methods for determination of luciferase activity vary greatly, therefore no specific method is described here. The lysis buffer used in these experiments is compatible with all methods tested, however, in specific cases other lysis buffers may be used. Repeated freeze–thaw cycles that are used in some methods should be avoided in these experiments, especially when luciferase activities are low, since freeze-thawing tends to decrease luciferase activity.

2. Microinjection in zebrafish has been described by others *(9)* and may be achieved in many ways. The method described here to mount embryos is very convenient but requires a (homemade) plastic mold. If this is not immediately available, embryos may be mounted in a small volume of embryo medium against the edge of an object glass placed in an empty Petri dish. Although simple, the visibility of the injections is hampered by the light-breaking effect of the medium surface.
3. The quality of DNA for microinjection should be routinely monitored. In general, standard preparation methods such as Qiagen columns or cesium chloride gradient centrifugation yield DNA of sufficiently high quality.
4. As mentioned previously, quantitative promoter analysis using luciferase should be preceded by spatial promoter analysis whenever possible to determine the optimal amount of DNA to be used for microinjection. When this can not be realized for some reason, it is important to vary the DNA concentration used for microinjection to determine the lowest possible amount of DNA that supports reliable detection of luciferase activity. Excessive amounts of DNA may result in nonspecific and ectopic activity, which will inevitably lead to the wrong conclusions.
5. The dissociation step in the dissociated blastula cell method is best practiced first on noninjected embryos. Spin down and examine the cells under a microscope. Viability may be checked with a standard dye exclusion test. The cells may be counted to check recovery, calculating the starting number from the estimated amount of cells in a single embryo. According to my experience, some of the cells (between 25% and 75%) are lost. In general, this should not present a problem, as long as enough cells are recovered to allow detection of luciferase activity. If no cell pellet or a very small pellet is observed after spinning down the dissociated blastula cells, the cells have probably been broken during trituration of the embryos. To prevent this, quickly suck the embryos through the needle and push the suspension out gently, avoiding air being blown into the liquid. Flick the tube to disperse the embryos that are still intact and repeat the procedure. Do not proceed beyond the point that all chorions are broken. If necessary, remaining aggregates may be dissociated by pipetting up and down after the chorion debris has been filtered off. Alternatively, embryos may be dechorionated manually and subsequently dissociated in PBS, although this is laborious and time-consuming.
6. When a lacZ expression vector is used for correction of luciferase data, high corrections (over fivefold) should be considered with caution. Either samples with very low lacZ activity should be left out of the analysis, or the experimental setup should be changed to decrease the variation.
7. Reporter genes other than luciferase and β-galactosidase may be used in promoter analysis experiments. Especially noteworthy is green fluorescent protein (GFP), which allows analysis in live embryos *(12)*. An increasing number of engineered GFPs has become available which are more stable, give stronger signals and allow faster detection than the original wild-type GFP (e.g., *13*). It should be noted however, that analysis of live embryos is often more laborious than of fixed and stained material. Especially if comparison of multiple (mosaic) embryos

is required to process and interpret the results, lacZ staining is preferred. However, analysis in live embryos allows the investigation of other aspects of promoter activity, such as cell movements. For quantitative analyses, luciferase is preferred, because detection is rapid, highly sensitive, and easy to perform.

8. As mentioned in **Note 10**, lacZ staining can be performed on embryos inside their chorion. Increase the length of the PBS washes after fixation to 10 min and proceed as usual. This is especially useful when large series are analyzed, of which only a small percentage is positive.
9. Murray's clearing solution is toxic and potentially carcinogenic and should therefore be handled with care. Wear gloves and work in properly ventilated rooms.
10. Promoter analysis using lacZ as a reporter may be combined with wholemount *in situ* hybridizations in the same embryo. To this end, the embryos are fixed for 45 min in ice-cold 4% paraformaldehyde in PBS and washed 10 min twice with PBS. LacZ staining is carried out overnight inside the chorions. Subsequently, embryos are postfixed for 2 h in 4% paraformaldehyde in PBS at room temperature, washed with PBS, dechorionated and processed further for the wholemount *in situ* hybridization. To achieve better contrast between the lacZ staining and the blue whole mount alkaline phosphatase staining, alternative β-galactosidase substrates are available from Molecular Probes that produce differently colored precipitates.
11. In some experiments, strong lacZ activity is observed in the yolk, which obscures staining in other regions of the embryo. This may be caused by leakage of DNA into the yolk during injection, which is transcribed in the yolk syncytial layer (YSL) nuclei. Try injecting in the center or in animal regions of the blastomeres and decreasing the injection volume. Take care to remove all injected embryos where leakage is observed. If this does not help, the promoter may be transported into YSL nuclei later during development. In that case, the yolk should be removed prior to inspection of the embryos.

Acknowledgments

I would like to thank Danica Zivkovic for critical comments on the manuscript. I am grateful to Sandra van de Water and Mark de Graaf who helped to develop part of the described methods.

References

1. Kriegler, M. (1990) Eukaryotic control elements, in *Gene Transfer and Expression*, Stockton, New York, pp. 3–21.
2. Gordon, J. W. (1993) Production of transgenic mice. *Methods Enzymol.* **225,** 747-771.
3. Sentry, J. W. and Kaiser, K. (1995) Progress in Drosophila genome manipulation. *Transgenic Res.* **4,** 155–162.
4. Mello, C. C., Kramer, J. M., Stinchcomb, D., and Ambros, V. (1991) Efficient gene transfer in *C. elegans*. *EMBO J.* **10,** 3539–3970.

5. Kroll, K. L. and Amaya, E. (1996) Transgenic *Xenopus* embryos from sperm nuclear transplantiations reveal FGF signaling requirements during gastrulation. *Development* **122,** 3173–3183.
6. Watabe, T., Candia, A., Rothbacher, U, Hasimoto, C., Inoue, K., and Cho, H. W. Y. (1995) Molecular mechanisms of Spemann's organizer formation: conserved growth factor synergy between *Xenopus* and mouse. *Genes Dev.* **9,** 3038–3050.
7. Gaiano, N. and Hopkins, N. (1996) Introducing genes into zebrafish. *Biochim Biophysica Acta* **1288,** 11–14.
8. Joore, J., Fasciana, C., Speksnijder, J. E. ., Kruijer, W., Destree, O. H. J., van den Eijnden-van Raay, A. J. M., et al. (1996) Regulation of the zebrafish goosecoid promoter by mesoderm inducing factors and Xwnt1. *Mech. Dev.* **55,** 3–18.
9. Westerfield, M. (1995) *The Zebrafish Book*, 3rd ed., University of Oregon Press, Eugene, OR.
10. Brasier, A. R., Tate, J. E., and Habener, J. F. (1989) Optimized use of the firefly luciferase assay as a reporter gene in mammalian cell lines. *Biotechniques* **7,** 1116–1122.
11. Westerfield, M. (1993) *The Zebrafish Book*, 1st ed., University of Oregon Press Eugene, OR.
12. Meng, A. M., Tang, H., Ong, B. A., Farrel, M. J., and Lin, S. (1997) Promoter analysis in living zebrafish embryos identifies a cis-acting motif required for neuronal expression of GATA-2. *Proc. Natl. Acad. Sci. USA* **94,** 6267–6272.
13. Zernicka-Goetz, M., Pines, J., Ryan, K., Siemering, K. R., Haseloff, J., Evans, M. J., et al. (1996) An indelible lineage marker for *Xenopus* using a mutated fluorescent protein. *Development* **122,** 3719–3724.

14

Transient Transgenesis in *Xenopus laevis* Facilitated by AAV-ITRs

Yuchang Fu, Donghui Kan, and Sylvia Evans

1. Introduction

A common approach to the study of gene function in *Xenopus* embryos is to ectopically express the gene of interest, or a mutant of that gene, by injecting either RNA or DNA, encoding the gene into early-stage embryos. Both of these approaches have disadvantages *(1)*. RNA injection results in little temporal or spatial control, as the RNA begins to be translated immediately following injection. In addition, RNA is unstable; therefore, transgene expression is relatively transient. In principle, DNA injections could ameliorate these problems. DNA does not begin to be translated until embryonic stage 8, at the mid-blastula transition, when zygotic genes begin to be expressed. Promoters can be utilized, which should enable more spatial and temporal control of gene expression, and DNA is more stable than RNA, prolonging the time frame in which transgenes can be expressed. However, for reasons that are not clear, expression of DNA plasmids following direct injection into embryos is extremely mosaic, and tissue specificity of promoters is often lost.

Recently, a method for the generation of stably transgenic frogs has been developed *(2)*, referred to as restriction enzyme-mediated integration (REMI). This method overcomes the shortcomings of transient transgene expression, enabling efficient and tissue-specific expression of DNAs following their integration into sperm DNA, and the subsequent fertilization of eggs by the transgenic sperm nuclei. This method, however, is technically very demanding.

We have recently developed an alternative, technically simpler, method for the expression of DNA transgenes in *Xenopus* embryos, which also results in an increased efficiency of transgene expression and, at least in the case of one promoter, tissue-specific transgene expression *(3)*. This method involves the construction of a vector DNA containing inverted terminal repeat sequences (ITRs) from an adeno-associated virus (AAV) *(4)* bracketing an expression cassette. The resulting constructs are directly injected into *Xenopus* embryos utilizing standard techniques (*see* **refs.** *1* and *5*; Chapters 10 and 12 of this volume).

We initially considered investigating the ability of AAV-ITRs to improve transgene expression in *Xenopus* following reports that plasmid DNAs containing these sequences (rAAV) exhibited enhanced and prolonged transgene expression in mammalian cells *(6)*. Adeno-associated virus is a nonautonomous parvovirus requiring coinfection with another virus for productive infection of cells and is a completely nonpathogenic integrating virus *(7)*. AAV vectors have been developed in which most of the viral genome is replaced by a transgenic expression cassette, retaining only the viral terminal repeat sequences of 145 nucleotides. The terminal 125 nucleotides of each ITR form palindromic hairpin structures that are required for DNA replication, integration, and excision of proviral sequences following integration. Although rAAV plasmids are capable of integration, enhanced transgene expression does not appear to require integration *(8)*.

Our results in *Xenopus* embryos have demonstrated that inclusion of AAV-ITRs on DNA plasmids can significantly increase transgene expression from a ubiquitous viral enhancer (*see* **Fig. 1**). With ITR constructs, up to 65% of transgenic embryos expressed the reporter transgene in more than 50% of descendents of injected cells, whereas only 16% of embryos showed comparable transgene expression when injected with non-ITR-containing plasmids. Up to 20% of ITR-plasmid-injected embryos expressed the transgene in 90–100% of the progeny of the injected cell, versus 3% with non-ITR control plasmids. These efficiencies of transgene expression are comparable to those obtained with REMI *(2)*.

Increased transgene expression does not appear to be a reflection of differences in DNA replication or stability, but rather appears to reflect, in part, improved segregation of plasmid DNA. Southern blot analysis provided little evidence that significant integration of transgene DNA was occurring *(3)*.

In addition to enhancing expression of a transgene driven by a ubiquitously expressed enhancer, the inclusion of ITRs enabled efficient and tissue-specific expression of a reporter transgene driven by a striated muscle specific promoter, that of the alpha cardiac actin gene *(9)*. Ectopic expression of the transgene outside of the domain of the endogenous gene was greatly reduced by the presence of the ITRs.

The ability of the ITRs to enhance both ubiquitous and tissue-specific transgene expression and to restrict non-tissue specific expression of the tissue-specific promoter is reminiscent of the ability of matrix attachment sites to overcome position effects *(10)*. In this regard, it is of interest to note that plas-

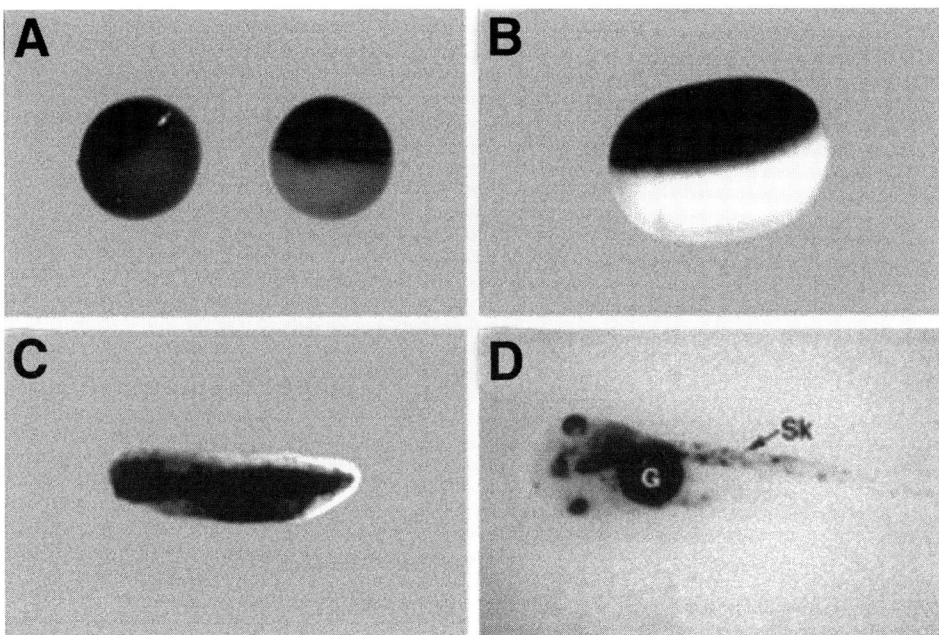

Fig. 1. X-gal staining of embryos injected with CMV $n\beta$-galactosidase. Embryos were injected with 200 pg of plasmid DNA into one cell at the two-cell stage, allowed to develop to the stages shown, fixed, and stained for β-galactosidase activity (dark) as described previously *(1)*. The embryos shown were scored as expressing in 90–100% of the descendants of the injected cell. (**A**) Stage-10 (left) and stage-12 (right) embryos. Arrow indicates rim of open blastopore. (**B**) Ventral view of a stage-16 neurula embryo, anterior to the left. (**C**) Lateral view of a stage-35 tailbud embryo; anterior to the left, dorsal at the top. (**D**) Dorsal view of a 1-mo-old tadpole, anterior to the left. G = gut, Sk = skeletal muscle. (Reproduced with permission from **ref. 3**.)

mids containing ITR sequences are localized in a perinuclear distribution in mammalian cells *(11)*. These observations are supportive of the idea that ITR sequences are capable of being recognized by a nuclear membrane-associated protein or factor. In mammalian cells, there is evidence that ITR sequences associate with numerous host-cell proteins *(12)*.

Our constructs were made utilizing one copy of the ITR from the left end of AAV and one copy of the ITR from the right end of AAV in an inverted repeat configuration. As this configuration was very effective, we have not tried other possible configurations. It is possible that diverse configurations, or multimerization of ITRs, could further improve transgene expression. It is also possible that other viral terminal repeat structures could improve transgene expression. These possibilities remain to be explored.

2. Materials

All solutions should be made from molecular-biology-grade reagents using sterile distilled water if not otherwise specified.

2.1. Construction of ITR-Containing Plasmids for Transgene Expression

1. Petri dishes and culture tubes for bacterial cell culture.
2. Eppendorf centrifuge.
3. Eppendorf tubes.
4. Restriction enzymes, ligase, and appropriate buffers (*see* manufacturer's instructions).
5. 37°C water bath.
6. Agarose gel electrophoresis apparatus: 1X Tris acetate EDTA (TAE) buffer: 40 mM Tris–acetate, 1 mM EDTA.
7. DNA plasmids: ITR-containing vector (*see* **Note 1**) and expression vector (e.g., CS2+).
8. Good rec- *E. coli* host strain (e.g., STABL2 cells, Gibco/BRL Gaithersburg, MD; *see* **Note 2**).

2.2. Microinjection of Xenopus Embryos

1. Male and female African clawed toads, *Xenopus laevis*. Albino animals should be used if wholemount *in situ* analyses will be performed. We purchase our frogs from Nasco Co. (Fort Atkinson, WI).
2. Human chorionic gonadotropin (HCG) (Sigma, St. Louis, MO, CG-10). Make up in sterile distilled water and store at 4°C.
3. 1X Marc's modified Ringer's (MMR) buffer: 100 mM NaCl, 2 mM CaCl$_2$, 2 mM KCl, 1 mM MgCl$_2$, 5 mM HEPES, pH 7.4.
4. 2% Cysteine hydrochloride solution for dejellying eggs prior to fertilization: always make fresh. For 100-mL solution, 2 g cysteine hydrochloride (Sigma C-7880), 10 mL of 1X MMR. Adjust pH between 7.98 and 8.1 with 10 N NaOH.
5. 3% Ficoll solution; 1.5 g Ficoll 400 (Pharmacia Biotech 17-0400-01) in 50 mL 0.5X MMR containing sodium penicillin and streptomycin sulfate (10 µg/mL, Gibco/BRL).
6. Microinjection apparatus: A number of injector systems can be used, described in some detail by Kay (*5*). We use a Drummond Nanojector (Broomall, PA) (cat. no. 3-00-203-X). Micropipets are held in a micromanipulator (we use a Narashige model M152 [Brinkmann Instruments, Westbury, NY]; again *see* **ref. 5** and Chapter 10 for other systems).
7. Dissecting Microscope and dual-fiber light pipes.
8. Reagents for X-gal staining or wholemount *in situ* analyses as previously described (**refs.** *1* and Chapters 5 and 12).

3. Methods

3.1. Construction of ITR-Containing Vectors and Preparation of DNA for Injection

1. Utilizing standard cloning techniques (*14*), ligate an ITR-containing fragment from appropriate source (*see* **Note 3**) to an expression cassette so that ITR

sequences bracket the expression cassette on either side. We routinely use the expression vector pCS2+ *(15)* as a backbone for our constructs (*see* **Note 4**).
2. Linearize the DNA template. We have found that linearized templates are more efficiently expressed than circular plasmids following injection into *Xenopus* embryos. Plasmids should be linearized so that the expression cassette flanked by the ITR sequences remains intact.
3. Following restriction digest, extract plasmid DNA with 1 volume of phenol–chloroform (1:1), and 1 volume of chloroform. Remove the upper aqueous phase, taking great care to avoid the organic phase, which is toxic to embryos. Add 1/10th volume sodium acetate, pH 5.2, and 2 volumes ethanol to precipitate.
4. Resuspend the DNA pellet in sterile distilled water. Measure the O.D. at 260 nm and calculate the DNA concentration. Confirm the DNA concentration by running a small aliquot on an agarose gel with a known amount of a DNA standard.

3.2. Microinjection of Plasmids into Xenopus *Embryos*

1. *Xenopus* embryos are obtained by established procedures (*16*; Chapter 10 of this volume). Briefly, ovulation is induced by injecting 300 U/frog of human chorionic gonadotropin into the dorsal lymph sac approximately 12 h prior to laying. Eggs are fertilized with sperm from macerated testes in 0.1X MMR. Fertilized eggs are dejellied in 2% cysteine HCl (pH 7.8) immediately following cortical rotation (after about 30 min at room temperature).
2. Microinjection of embryos is performed according to standard protocols (*5*; Chapter 10 of this volume). Briefly, load microinjection needles with DNA samples by air or oil. We routinely inject 10 to 20 nL of DNA sample (10–20 pg/nL; *see* **Note 5**) into one cell of a two-cell stage embryo, as viability is improved in comparison to injections at the one-cell stage, in addition, the noninjected side can serve as an internal control for the experiment. If required, targeted injections into specific blastomeres can be performed using 4–32 cell embryos (*see* Chapter 12).
3. Inject embryos in 0.5X MMR + 3% Ficoll 400 solution, and following injection are keep at 15–18°C. After several cell cycles (4–6 h), transfer the embryos to 0.2X MMR solution with 1% Ficoll 400, sodium penicillin, and streptomycin sulfate (10 µg/mL) for further development.
4. Developmental stages of the injected embryos are determined according to Nieuwkoop and Faber *(17)*.
5. Score for transgene expression (*see* **Note 6**): Expression of a β-galactosidase reporter transgene can be scored by X-gal staining of embryos (*1*, *see* Chapter 12) or by wholemount *in situ* analysis for lacZ mRNA or immunohistochemistry of the protein (*13*; Chapters 5 and 8, of this volume).

4. Notes

1. The high degree of internal secondary structure present within the ITRs prevents their being readily sequenced by the dideoxy-termination method and also prevents analysis by polymerase chain reaction. To monitor for the presence of intact

ITRs, we use restriction analysis with SmaI and BglI enzymes. Within the ITR sequence, there are two SmaI sites and one BglI site within a region, which, in our experience, is highly susceptible to deletion or rearrangement (SmaI is methylation sensitive; BglI is not). The complete adeno-associated virus 2 DNA sequence, including ITR sequences, can be found in GenBank, accession numbers J01901, M12405, M12468, and M12469.
2. AAV-ITR sequences can be very unstable, frequently undergoing deletion or rearrangement. Therefore, it is of utmost importance to propagate ITR-containing plasmids in a reliable rec- *E. coli* host strain. *E. coli* SURE2 cells from Stratagene have been recommended by Nick Muzyczka at the University of Florida, and we have been using STABL2 cells from Gibco/BRL. Even in these rec- strains, it is important to monitor each DNA preparation for deletion of ITR sequences. It is important to isolate several individual colonies from each transformation.
3. Source of the AAV-ITRs. Owing to the instability of the ITRs, it is important to obtain an ITR-containing plasmid with intact ITRs.
4. The pCS2+ expression vector is designed to be expressed efficiently in both *Xenopus* and mammalian cells, it contains the strong and ubiquitously expressed cytomegalovirus (CMV) enhancer, 5' untranslated regions from the β-globin gene that allow for efficient translation, a multiple-cloning site, and a polyadenylation signal from SV40. The CMV enhancer can be excised and replaced by a tissue-specific promoter. A CS2 vector which contains nuclear localized β-galactosidase, which is readily detected in frog embryos, is available from Dave Turner, Department of Neurosciences, University of Michigan, Ann Arbor, MI.
5. It is important to perform titrations of each DNA preparation to be injected to empirically determine the optimum amount for achieving a balance between good expression and embryonic lethality.
6. Transgene expression is very dependent on the quality of the embryos, which can vary widely depending on egg quality, sperm quality, and the season.

References

1. Vize, P.D., Hemmati-Brivanlou, A., Harland, R. M., and Melton, D. A. (1991) Assays for gene function in developing embryos, in *Methods in Cell Biology*, Vol. 36 (Kay, B. K. and Peng H. B., eds.), Academic Press, pp. 368–387.
2. Kroll, K. L. and Amaya, E. (1996) Transgenic *Xenopus* embryos from sperm nuclear transplantations reveal FGF signaling requirements during gastrulation. *Development* **122,** 3173–3183.
3. Fu, Y., Wang, Y., and Evans, S. (1998) Viral sequences enable efficient and tissue-specific expression of transgenes in *Xenopus*. *Nature Biotech.* **16,** 253–257.
4. Lusby, E., Fife, K. H., and Berns, K. I. (1980) Nucleotide sequence of the inverted terminal repetition in adeno-associated virus DNA. *J. Virol.* **34,** 402–409.
5. Kay, B. (1991) Injection of oocytes and embryos, in Methods in Cell Biology (Kay, B. K. and Peng, H. B., eds.), Academic Press, **36,** 663–669.
6. Philip, R. , Brunette, E., Kilinski, L., Murugesh, D., McNally, M.A., Ucar, K., et al. (1994). *Mol. Cell. Biol.* **14,** 2411–2418.

7. Lebkowski, J. S., McNally, M. M., Okarma, T. B., and Lerch, L. B. (1988). Adeno-associated virus:a vector system for efficient introduction and integration of DNA into a variety of mammalian cell types. *Mol. Cell. Biol.* **8,** 3988–3996.
8. Flotte, T. R., Afione, S. A., and Zeitlin, P. A. (1994) Adeno-associated virus vector gene expression occurs in nondividing cells in the absence of vector DNA integration. *Am. J. Respir. Cell. Mol. Biol.* **11,** 517–521.
9. Mohun, T. J., Garrett, N., and Gurdon, J. B. (1986) The CArG promoter sequence is necessary for muscle-specific transcription of the cardiac actin gene in *Xenopus* embryos. *EMBO J.* **8(4),**1153–1161.
10. Porter, S. D. and Meyer, C. J. (1994) A distal tyrosinase upstream element stimulates gene expression in neural-crest-derived melanocytes of transgenic mice: position-independent and mosaic expression. *Development* **120,** 2103–2111.
11. Weitzman, M. D., Fisher, K. J., and Wilson, J. M. (1996) Recruitment of wild-type and recombinant adeno-associated virus into adenovirus replication centres. *J. Virol.* **70,** 1845–1854.
12. Harmonat, P.L., Santin, A.D., Carter, C.A., Parham, G.P., and Quirk, J.G. (1997) Multiple cellular proteins are recognized by the adeno-associated virus Rep78 major regulatory protein and the amino-half of Rep78 is required for many of these interactions. *Biochem. Mol. Biol. Int.* **43(2),** 409–420.
13. Harland, R. M. (1991) In situ hybridization: an improved whole-mounve transgene expression in *Xenopus* method for *Xenopus* embryos, in *Methods in Cell Biology*, Vol. 36 (Kay, B. K. and Peng H. B., eds.), Academic Press, pp. 685–695.
14. Ausubel, F.M. (1992) *Current Protocols in Molecular Biology.* Greene Pub. Associates/Wiley-Interscience, New York.
15. Turner, D. L. and Weintraub, H. (1994) Expression of achaete-scute homolog 3 in *Xenopus* embryos converts ectodermal cells to a neural fate. *Genes Dev.* **8(12),** 1434–1447.
16. Sive, H. (1998) *Early Development of* Xenopus laevis*: A Course Manual*, Cold Spring Harbor Laboratory Press, Cold Spring Harbor, NY.
17. Nieuwkoop, P., and Faber, J. (1956) *Normal Table of* Xenopus laevis (Daudin). North-Holland, Amsterdam.

15

Band-Shift Analysis Using Crude Oocyte and Embryo Extracts from *Xenopus laevis*

Rob Orford and Matt Guille

1. Introduction

The DNA-binding proteins play a pivotal role in the cell, regulating gene transcription, DNA replication and repair, it is therefore of fundamental importance that the mechanisms controlling these factors and their interactions be investigated. The study of DNA-binding proteins in the developing embryo would be extremely limited if one did not consider the use of the electrophoretic mobility shift assay, (EMSA also known as the gel retardation or band-shift assay) *(1,2)*. Gel retardation is a fast, reliable and inexpensive assay, which can provide a wealth of information about DNA-binding proteins *(3–5)*. Band-shift analysis also provides an entry point to understanding the nature of DNA–protein interactions within the embryo.

The DNA–protein interactions within the cell are governed by multiple factors, which, overall, represent a very complex environment. EMSA allows the investigator to unmask the complexities of such interactions by considering each binding parameter separately in an attempt to reconstruct the in vivo situation *(6)*.

As our main example, we will consider the study of transcription factors using this assay, although similar methods have been applied to other types of DNA-binding proteins *(7)*. Using band-shift analysis as a semiquantitative tool, it is possible to determine the presence of a transcription factor within the embryo during development (*see* **Fig 1**) *(1,8,9)*. Gene expression may depend not only on the presence of the transcription factor, within the embryo, but also on a plethora of other components such as cofactors, ligands, metal ions, chromatin structure, and subcellular localization. Ascertaining the identity of such

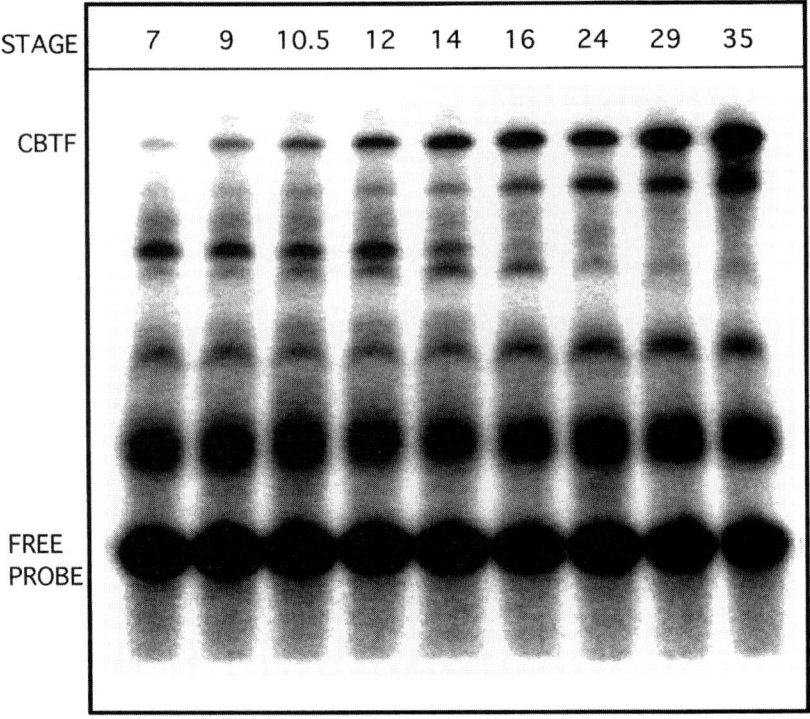

Fig. 1. CCAAT-Box transcription factor (CBTF) activity during early embryonic development. Batches of 15 embryos were harvested and snap-frozen at each stage. After homogenization and Freon extraction, the equivalent to half an embryo was used in each reaction with a 4-fmol wild-type end-labeled probe, 500 ng poly(dI-dC)–(dI-dC) and 4% Ficoll. The free and bound DNA were then separated on a 4% polyacrylaminde, 0.25X TBE gel; the gel was then dried and underwent autoradiography.

regulating elements is relatively straightforward using this assay. The bandshift assay allows the investigator to identify which of these components (excluding chromatin and subcellular localization) affect formation of the DNA–protein complex. Other types of study can be carried out using gel retardation. The identity of transcription factor subunits may be resolved by the inclusion of specific antibodies in the assay *(10,11)*. The nature of the protein–DNA interaction can be analyzed using gel retardation in conjunction with DNA-footprinting techniques (*see* Chapter 16) *(4)*. Band-shift analysis is also an extremely useful technique when purifying a transcription factor from embryo or oocyte extracts, the assay allows the investigator to monitor the success of purification steps and provides an insight into how much physical and

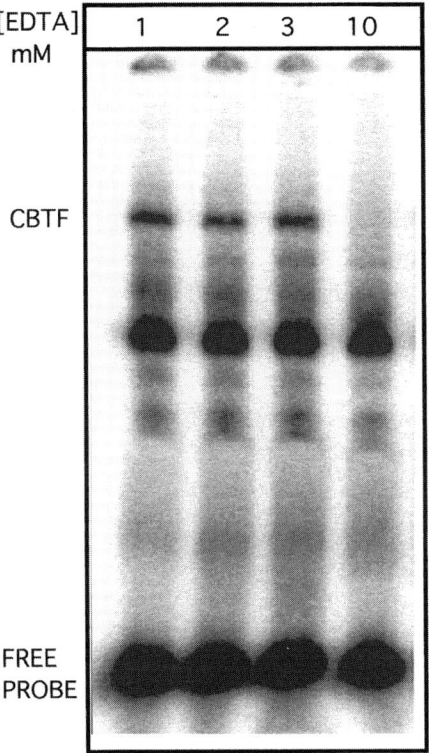

Fig. 2. CBTF remains active in the presence of EDTA at concentrations of 3 mM. Oocytes were taken from a single female *Xenopus*. After homogenization and Freon extraction, the equivalent to half an oocyte was used in each reaction; it was exposed to the EDTA concentrations shown for 30 min at 4°C prior to incubation with a 4-fmol wild-type end-labeled probe, 500 ng poly(dI-dC)–(dI-dC), and 4% Ficoll. The free and bound DNA were then separated on a 4% polyacrylaminde, 0.25X TBE gel; the gel was then dried and underwent autoradiography.

chemical abuse the transcription factor is able to withstand and remain active (*see* **Note 1** and **Fig. 2**) *(12–14)*.

To perform the band-shift assay the investigator must first obtain an oligonucleotide (wild-type probe) or restriction fragment that contains a specific recognition sequence for a transcription factor, whose binding in vivo leads to activation or suppression of a downstream gene. A second oligonucleotide must be obtained (mutant probe), whereby a mutation has been incorporated into the recognition sequence, which, when introduced into the regulatory sequence in vivo, abolishes expression of the target gene (*see* **Note 2**). To carry out the assay, the wild-type DNA probe is radioactively labeled, and incubated with homogenized embryo extract in the presence of excess nonspecific competitor DNA. The inclusion of an excess of nonspecific competitor will increase the sensitivity of the assay, provided that the DNA-binding protein in question has a much higher affinity for the wild-type probe than the nonspecific competitor (*see* **Note 3**) *(15)*. This reaction mix is then subjected to electrophoresis on a native polyacrylamide gel. The free DNA probe carrying a net negative charge will migrate through the gel matrix at the highest rate, however DNA that is complexed to protein will migrate at a slower rate. Exposing the protein–DNA complex to a polyacrylamide gel matrix effectively "freezes" the dynamic equilibrium that would be observed during the electrophoretic dead-time (*see* **Note 4**). The gel then undergoes autoradiography, the resulting autoradiograph will almost certainly contain a number of bands which have been retarded. Determining which of these bands represent the specific protein–DNA complex under investigation is straightforward, by adding approx 100- to 1000-fold excess of unlabeled (cold) wild-type probe to the original reaction mix, a band representing that of the specific complex should disappear. Addition of an excess of cold mutant probe should not affect the intensity of the specific band. A decrease in nonspecific band intensity will occur in the presence of wild-type or mutant competitors. By altering the binding conditions of the reaction mix and subsequently monitoring the increase or decrease of the specific band, optimum reaction conditions can be established (*see* **Fig. 3**).

A band-shift reaction can be modified to aid in the identification of the protein(s) bound to a specific DNA sequence provided that an antibody to that protein is available. The inclusion of an antibody specific for a component of the protein complex in the band-shift reaction can result in further retardation (supershift) of the band by decreasing the electrophoretic mobility of the complex. A second possibility is encountered when the epitope for the antibody is within the DNA-binding domain of the transcription factor. A modified supershift (blocking assay) can be performed such that the DNA-binding activity of the protein is eliminated, because of the steric hindrance of the bound antibody. This is achieved by the addition of the antibody prior to the addition of probe.

Using the band-shift assay, it is occasionally possible to monitor the temporal regulation of transcription factors within the nucleus. The technique of band-shifting nuclear embryo extracts is difficult, particularly when early embryos are involved. However, such labors can yield much information concerning the regulation of cytoplasmic to nuclear translocation of transcription factors, which is clearly a major mechanism of transcriptional control in early development.

2. Materials

1. 10X tris-borate EDTA (TBE): 0.89 M Tris-borate pH 8.3, 20 mM EDTA, electrophoretic sequencing grade (National Diagnostics, Hull, UK).

Band-Shift Analysis of Oocyte and Embryo 179

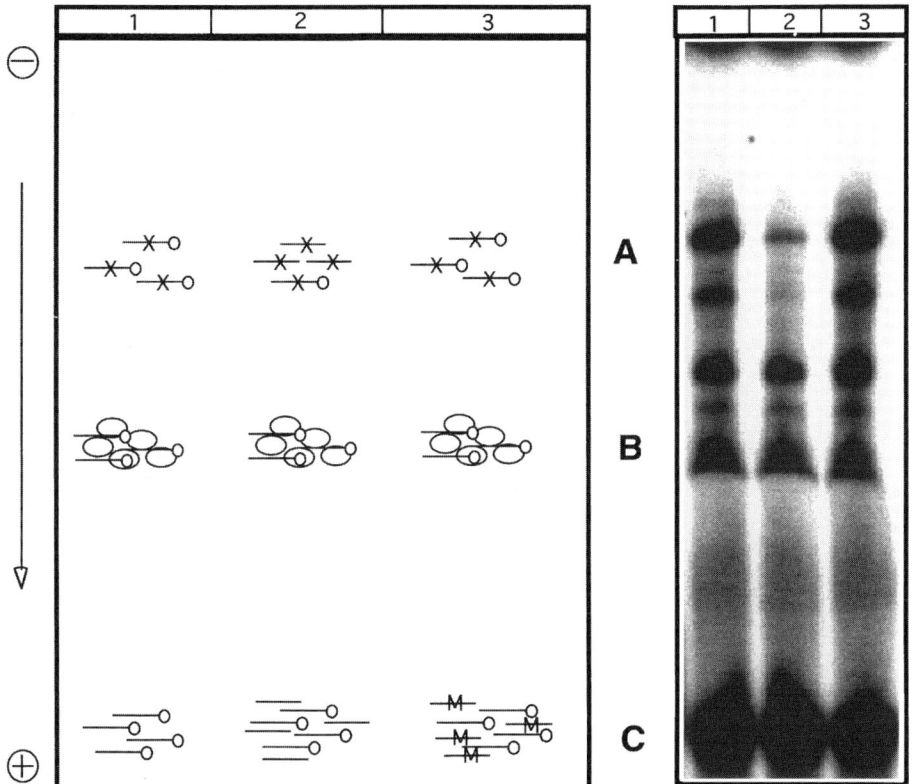

Fig. 3. The band-shift gel represents a specific and nonspecific competition. Lane 1 contains a 4-fmol 5' radiolabeled wild-type (WT) probe (—O) and 5 µL of crude embryo extract. Lane 2 contains the same amount end-labeled WT probe and embryo extract, plus 100-fold molar excess of unlabeled WT probe (—). Lane 3 is essentially the same as lane 2 with the 100-fold excess WT probe replaced with the same quantity of mutant (MT) probe (M). Position A represents specific protein–DNA complex (-×O) identified by the decrease in intensity with excess WT probe (lane 2) and unchanged with excess MT probe (lane 3). Position B represents probe that has been trapped in the cellular debris interface (O) migrating through the gel. Position C represents the free probe.

2. Acrylamide sequencing grade 40% (29:1) acrylamide: bisacrylamide solution (Gas stabilized) (Accugel 29:1 ultra pure, National Diagnostics).
3. TEMED (N,N,N',N',-tetramethylethylenediamine) electrophoresis grade (National Diagnostics).
4. Ammonium persulfate (APS): 10% (w/v) in water (Sigma, Poole, UK), store at 4°C for <1 mo.
5. Vertical electrophoresis system (e.g., Atto (2 mm thickness, 12 space) and power supply capable of 200 V and 40 mA.

6. Vacuum gel drier
7. EMSA extraction buffer: 20 mM HEPES (pH 7.9), 2 mM MgCl$_2$ 10 mM β-glycerophosphate (Sigma), protease inhibitor cocktail tablets, Complete™ Mini, EDTA-free (Boehringer Mannheim, Lewes, UK), made fresh.
8. EMSA reaction buffer: 4% (w/v) Ficoll, 20 mM HEPES (pH 7.9), 2 mM MgCl$_2$, 50 mM NaCl (Sigma) store at 4°C.
9. 1,1,2-Trichlorotrifluoroethane (Freon) (Sigma).
10. 3MM Chromatography paper (Whatman, Maidstone, UK).
11. Poly(dI-dC)–(dI-dC) rehydrated to 1 mg/mL in neutral buffer containing 100 mM NaCl and stored at –20°C (Pharmacia, St. Albans, UK).
12. T4 Polynucleotide kinase 10,000 U/mL (NEB, Hitchin, UK).
13. 10X polynucleotide kinase (PNK) buffer (0.5 M Tris HCl, pH 7.6, 0.1 M MgCl$_2$, 0.05 M dithiothreitol [DTT]) (NEB).
14. γ-[^{32}P]–ATP 9.25 Mbq, 10 µCi/µL (Amersham, Little Chalfont, UK).
15. G25 Sepharose (fine matrix) (Pharmacia).
16. Spin-X tubes (Sigma).
17. Tris-sodium chloride EDTA (TNE) buffer: Tris, 100 mM NaCl, 2 mM EDTA.
18. Nonidet 40 (NP-40) (Merck, Lutterworth, UK).
19. Refrigerated bench-top centrifuge (variable speed, capable of 14,000g).
20. X-ray film and developing facilities.
21. Autoradiography cassette.
22. Adult female *Xenopus laevis*.
23. 0.1X modified Barth's saline (MBS), 150 mg/mL cycloheximide (*Caution*: toxic).
24. Homogenization buffer containing 2% sucrose.
25. Homogenization buffer containing 0.8 M sucrose.
26. High salt buffer (HSB): homogenization buffer containing 300 mM KCl.
27. Embryo manipulation facilities, stereoscope (e.g., Nikon, SMZ-U), cold-light source (e.g., Schott KL1500).
28. Cold room or cold cabinet.

3. Methods
3.1. Standard Band-Shift Protocol

The band-shift protocol is relatively straightforward; however, assaying one specific transcription factor in a crude embryo extract can be difficult. The embryo preparation and binding reactions need careful planning, as unpurified transcription factors are almost invariably labile. Therefore reagents should be prepared and centrifuges precooled prior to preparing embryo extracts.

3.1.1. Probe Preparation

1. Dissolve the oligonucleotides in TNE buffer and store at –20°C.
2. Prior to commencing the labeling reaction, dilute the oligonucleotides to 200 fmol/µL.

3. Anneal equal volumes of both strands. Heat to 95°C in a hot block for a few minutes, switch off and leave to cool overnight.
4. 5' End-label the dsDNA probe as follows: to a 1.5-mL Eppendorf tube, add 2 µL of 200 fmol/µL dsDNA probe, 1 µL 10X PNK buffer, 30–50 µCi γ-[^{32}P]-ATP, make up to 9 µL with dH$_2$O, and finally add 1 µL T4 PNK.
5. Incubate at 37°C for 30 min.
6. During the incubation period, add 0.75 mL of G25 Sepharose (Sigma) suspension (the matrix is usually stored at 4°C in TNE buffer) to a spin-X tube, remove the liquid from the matrix by centrifuging for 30 s at full speed in a microcentrifuge.
7. After the incubation period, add 90 µL TNE buffer to the labeling reaction.
8. Remove the unincorporated radioisotope from the reaction by spinning the diluted reaction through the G25 in the spin-X tube for 30 s at full speed in a microcentrifuge (*see* **Note 5**).

3.1.2. Preparation of EMSA Gel and Reaction Mix

1. Assemble the electrophoresis plates, ensuring that the rubber gasket is secure so that the plates do not leak. Prepare a native 0.25X TBE, 4% polyacrylamide gel as follows: To a beaker, add 1.25 mL 10X TBE buffer, 5 mL 40% (29:1) acrylamide:bisacrylamide solution, 43.3 mL dH$_2$O, and finally, add 400 µL APS and 50 µL TEMED. The addition of TEMED acts to catalyze the production of free radicals by APS, this, in turn, polymerizes the acrylamide.
2. Mix the solution by inverting, pour into the assembled plates, and add the well-forming comb carefully to avoid bubble formation in the gel.
3. Allow the gel to set for 30 min at room temperature, cover with plastic wrap, and place at 4°C overnight.
4. Prior to performing the band-shift reaction, it is important to pre-run the polyacrylamide gel in the cold room for 2 h at 200 V (*see* **Note 6**), note that the current remains at constant when the gel has prerun for a sufficient period of time (approx 12 mA). Optimum binding conditions are determined by varying the concentration of one reactant at a time (*see* **Fig. 2**). The following binding reaction should only be considered as a guideline. To a chilled 1.5-mL Eppendorf tube, add 4 µL EMSA reaction buffer, 1 µL dsDNA probe (4 fmol/µL), 50 ng–1 µg poly(dI-dC)–poly(dI-dC), dH$_2$O to final volume of 10 µL (*see* **Note 7**).
5. Embryos or oocytes which have previously been collected and stored at −70°C should be defrosted by the addition of EMSA extraction buffer (5 µL per embryo) (*see* **Note 8**.)
6. Homogenize the embryos with a 1-mL pipet tip.
7. Add an equal volume of Freon and vortex vigorously for 30 s.
8. Centrifuge at 14,000*g*, 4°C for 3 min.
9. Decant the supernatant into a fresh 1.5-mL Eppendorf. The Freon extraction is necessary to remove lipoproteins, which can have a detrimental effect on the binding reaction, contribute to overloading the gel, and give false positives in immunological assays *(16)*.

10. Pipet 5 µL of the embryo extract into the 10 µL of solution prepared in **step 4**, vortex, and spin briefly.
11. Allow binding to occur for 15–30 min at 4°C.
12. After the binding period, flush each well with buffer and load the samples. Run the gel at 200 V for 95 min, so that the free or unbound probe (30–40 mer) remains on the gel.
13. Following the electrophoresis, remove the back plate from the gel and cut the left corner to facilitate orientation. Lift off the gel with 3 MM chromatography paper and cover with plastic wrap, then lay the gel on a second sheet of paper to avoid radioactive contamination of drying equipment.
14. Dry the gel on a vacuum drier for 1 h at 70–80°C.
15. Finally, autoradiograph the gel overnight or place on a Phosphorimager screen for a few hours if quantitation is required.

3.2. Supershift and Blocking Assay

The inclusion of an antibody in the band-shift assay can lead to further retardation or elimination of a shifted complex if the appropriate epitope is within the DNA–protein complex. Providing that the antibody has sufficient affinity and recognizes an epitope that is outside a critical binding region, then a supershifted band is observed, hence identifying the protein involved in the interaction. The addition of an antibody that binds to the DNA-binding domain of the transcription factor can interfere with the protein–DNA complex formation, which, in turn, results in the disappearance of the band of interest on the final autoradiograph. Bands further retarded or ablated by the addition of antibody are identified by comparison of controls using preimmune serum.

1. Assemble a standard binding reaction (as in **Subheading 3.1.2.**) and chill on ice.
2. Homogenize the embryos and prepare extract (as in **Subheading 3.1.2.**) (*see* **Note 9**).
3. To perform a blocking assay add the antibody to 5 µL of extract, initially using a range of volumes or concentrations. Preimmune serum is used in parallel, this is to ensure that any shifted or missing bands are specifically the result of the presence of the antibody alone.
4. Ideal binding temperatures differ for each protein; antibodies will generally form complexes with a greater efficiency at room temperature. However; these warmer conditions often result in degradation or inactivation of unpurified transcription factors. Allow 1 h for antibody binding.
5. Add the chilled binding reaction mix (from **step 1**) to the protein–antibody solution.
6. Allow the DNA to bind for 15–30 min at 4°C. The supershift differs from this assay in that the DNA is bound to the transcription factor prior to the addition of the antibody (i.e., **steps 5** and **6** are carried out prior to **steps 3** and **4**).
7. After binding, load and run the gel as described for the standard band-shift.

3.3. Preparation of Nuclear Extracts from Early Xenopus embryos

This method *(1)* is a modification of the method of Lemaitre et al. *(17)*. Briefly embryos are incubated in cycloheximide to block the rapid cell cycles occurring early in development and, hence, ensure that a nuclear membrane is intact. The embryos are broken up gently and centrifuged to pellet pigment granules and allow separation of the lipids present. The supernatant containing the nuclei is then centrifuged above a sucrose cushion that only the nuclei penetrate. The nuclei are then washed and lysed. When performing this procedure, it is essential to monitor the recovery of the nuclear material. We found that nuclear levels of maternal GATA-2 did not vary during early development *(1)*; hence, analyzing GATA-2 activity by gel shift in the samples provides a good control.

1. Incubate a set of 50–100 embryos in 0.1X MBS containing 150 mg/mL cycloheximide for 1 h prior to their reaching the required stage. This will need to be timed with the use of the Normal table and by incubation at 23°C.
2. Transfer the embryos to an Eppendorf tube, wash with homogenization buffer, and remove all liquid carefully using a Finn-type pipet.
3. Crush the embryos gently (we use a sealed Pasteur pipet).
4. Centrifuge at 10,000g for 10 min at 4°C.
5. Remove the supernatant; take care to avoid the lipid layer at the top.
6. Dilute the supernatant fivefold with homogenization buffer containing 2% sucrose.
7. Layer this over a cushion of homogenization buffer containing 0.8 M sucrose (0.5X the volume of the dilute supernatant) in an Eppendorf tube.
8. Centrifuge for 5 min at 4000g and 4°C.
9. Remove the supernatant.
10. Wash the pellet very gently in homogenization buffer.
11. Resuspend the pellet in HSB (0.5 mL per embryo).
12. Lyse the nuclei by rapid freeze–thaw.
13. Add four volumes of homogenization buffer and mix.
14. Pellet debris by centrifugation for 5 min at 10,000g and 4°C.
15. 10 µL of the nuclear lysate is usually adequate to give a band overnight on a gel shift.

4. Notes

1. During purification steps, samples are often exposed to different environments, which are dependent on the procedure used; for example, often DNA–affinity chromatography is utilized when purifying transcription factors from cell extracts. During DNA–affinity chromatography, it is important to include EDTA in buffers, as this eliminates the action of nucleases that require divalent cations as cofactors. Using band-shift analysis, it is possible to determine if EDTA will have any effect on the protein activity (*see* **Fig. 2**).
2. The length of DNA fragment to be used during the band-shift assay must be determined experimentally, as it is constrained not only by the minimal protein-

binding sequence but also by the resolvability of the specific complexes within the band-shift gel (*see* **Note 6**). The minimum length of DNA is generally considered to have at least 20 bases either side of the binding site, as this allows for further in situ analysis of the protein–DNA interaction by various footprinting and interference techniques (*see* Chapter 16). Small probes (under 100 basepairs) are readily obtainable as synthetic oligonucleotides. Gel purification of DNA fragments is also required. Gel purified oligonucleotides are obtainable from most oligo-producing companies. Gel purification can be achieved by ultraviolet shadowing the DNA after separation on a high-percentage polyacrylamide gel and eluting the requisite band. Larger DNA probes are usually prepared by restriction digests of plasmids containing the complete binding site, or a promoter fragment as an insert

3. The use of nonspecific competitors are essential when performing gel retardation on crude cell extracts. The inclusion of agents such as *E. coli* DNA, calf thymus DNA, heparin, or copolymers like poly(dI-dC)–poly(dI-dC) and poly(dG-dC)–poly(dG-dC) act to increase the sensitivity of the assay. The optimum quantities to be used must be determined experimentally. An excessive amount of nonspecific competitor will compete with the wild-type probe for transcription factor binding, which will then decrease sensitivity of the assay *(4)*. The simple alternating copolymers are most frequently used as they compete for sequence-specific DNA–binding proteins less effectively than the heterologous sequence DNA.

4. The action of the polyacrylamide gel is considered to cage the species trapped within the gel. The "freezing" of equilibrium refers to the progression of the reaction not changing after the electrophoretic dead time has passed. If the stoichiometry of the reaction is $aP+bD(cPD$ (where P is free protein, D represents DNA, and PD is the protein–DNA complex) at the last moment prior to entering the gel matrix, then this situation will be represented on the final autoradiograph.

5. Check that the DNA probe is sufficiently labeled by Cerenkov counting 1 µL of probe, 1500–2000 cps is considered acceptable. Avoid storing the radiolabeled DNA for periods longer than two weeks.

6. Prerunning polyacrylamide gels is often necessary to allow for free radicals produced during the polymerization reaction to dissipate. The presence of free radicals can have a detrimental effect on the DNA–protein complex, often eliminating binding. Prerunning the gel also ensures that the buffer is uniformly distributed and the gel is at a constant temperature.

7. The loading dye need not be added, as the sample contains Ficoll; this also eliminates band distortion, which is often observed if glycerol is used to increase the sample density.

8. Embryo preparation should be carried out on ice because unpurified transcription factors are generally immensely unstable, half-lives of a few minutes are not uncommon in crude extracts. Extract equivalent to half of an embryo (approx 50 µg total protein) is generally considered sufficient for each band-shift reaction.

Although the binding activity of the transcription factor dictates the amount of protein used in the band-shift reaction, total protein should be taken into account, as it is possible to overload the polyacrylamide gel.
9. The inclusion of nonionic detergents such as approx 0.1% (w/v) Nonidet 40 (NP-40) may act to reveal partially buried epitopes, without denaturing the transcription factor complex.

References

1. Brewer, A. C., Guille, M. J., Fear, D. J., Partington, G. A., and Patient, R. K. (1995) Nuclear translocation of a maternal CCAAT factor at the start of gastrulation activates *Xenopus* GATA-2 transcription. *EMBO J.* **14(4),** 757–766.
2. Huang, H., Murtaugh, L. C., Vize, P. D., and Whitman, M. (1995) Identification of a potential regulator of early transcriptional responses to mesoderm inducers in the frog embryo. *EMBO J.* **14,** 5965–5973.
3. Fried, M. G. and Crothers, D. M. (1981) Equilibria and kinetics of lac repressor–operator interactions by polyacrylamide gel electrophoresis. *Nucleic Acids Res.* **9,** 6505–6525.
4. Revzin, A. (1987) Gel electrophoresis assays for DNA–protein interactions. *Biotechniques* **7,** 346–355.
5. Ceglarek, J. A. and Revzin, A. (1989) Studies of DNA–protein interactions by gel electrophoresis. *Electrophoresis* **10,** 360–365.
6. Dent, C. L. and Latchman, D. S. (1993) *Transcription Factors, A Practical Approach* (Latchman, D. S., ed.), Oxford University Press, New York, pp. 1–26.
7. Mernagh, D. R., Reynolds, L. A., and Kneale, G. G. (1997) DNA binding and subunit interactions in the type I methyltransferase M.*Eco*R124I. *Nucleic Acids Res.* **25,** 987–991.
8. Mohun, T. J.,Garrett, N., and Gurdon, J. B. (1986) Upstream sequences required for tissue-specific activation of the cardiac actin gene in *Xenopus laevis* embryos. *EMBO J.* **5,** 3185–3193.
9. Snape, A. M., Jonas, E. A., and Sargent, T. D. (1990) KTF–1, a transcriptional activator of *Xenopus* embryonic keratin expression. *Development* **109,** 157–165.
10. Snape, A. M., Winning, R. S., and Sargent, T. D. (1991) Transcription factor AP-2 is tissue-specific and is closely related or identical to Keratin Transcription Factor 1 (KTF-1).*Development* **113,** 283–293.
11. Orford, R. L., Robinson, C., Haydon, J., Patient, R. K., and Guille, M. J. (1998) The maternal CCAAT box transcription factor which controls GATA-2 expression is novel, developmentally regulated and contains a dsRNA-binding subunit. *Mol. Cell. Biol.* **18,** 5557–5566.
12. Corthesy, B. and Kao, P. N. (1994) Purification by DNA affinity-chromatography of 2 polypeptides that contact the NF-AT DNA-binding site in the interleukin-2 promoter. *J. Biol. Chem.* **269,** 20,682–20,690.
13. Perkins, N. D., Nicolas, R. H., Plumb, M. A., and Goodwin, G. H. (1989) The purification of an erythroid protein which binds to enhancer and promoter elements of haemoglobin genes. *Nucleic Acids Res.* **17,** 1299–1314.

14. Treisman, R. (1987) Identification and purification of a polypeptide that binds to the c-fos serum response element. *EMBO J.* **6,** 2711–2717.
15. Sing, H., Sen, R., Baltimore, D., and Sharp, P. A. (1986) A nuclear factor that binds to a conserved sequence motif in transcriptional control elements of immunoglobulin genes. *Nature* **319,** 154–158.
16. Wyllie, A. H., Laskey, R. A., Finch, J., and Gurdon, J. B. (1978) Selective DNA conservation and chromatin assembly after injection of SV40 DNA into *Xenopus* oocytes. *Dev. Biol.* **64,** 178–188.
17. Lemaitre, J. M., Bocquet, S., Buckle, R., and Mechali, M. (1995) Selective and rapid nuclear translocation of a c-myc–containing complex after fertilization of *Xenopus laevis* eggs. *Mol. Cell Biol.* **15,** 5054–5062.

16

DNA-Footprinting using Crude Embryonic Extracts from *Xenopus laevis*

Rob Orford, Darren Mernagh, and Matt Guille

1. Introduction

The interactions of trans-acting factors with cis-acting DNA elements have been a mainstay of molecular biology. With the emergence of new techniques, it is becoming easier to investigate how specific genes are regulated in differentiating cells. In 1978, Galas and Schmitz developed a modified employment of the enzyme deoxyribonuclease I (DNase I) that made it possible to map the DNA-binding site of a putative transcription factor *(1)*. With the advent of Maxam and Gilbert sequencing, it also became apparent that chemical nucleases could be used to investigate the nature of protein–DNA interactions *(2)*. Although these techniques are applicable to other types of DNA-binding proteins, we will specifically consider transcription factors.

The technique of DNA footprinting originally involved exposing a purified DNA–protein complex to a chemical or enzymatic nuclease. The unprotected regions were nicked, or modified for later cleavage, on average once per DNA fragment. Protected regions of DNA were observed on a denaturing sequencing gel by comparison to cleaved, unbound DNA (*see* **Fig. 1**). The introduction of new chemical and enzymatic nucleases has provided the investigator with tools to analyze the specific areas of DNA contacted, or in close proximity to, the transcription factor (*see* **Fig. 1**) *(4)*. Such techniques have given information about DNA–protein interactions that had only previously been observed with X-ray crystallography. Unfortunately, early protocols worked only if the transcription factor in question was purified.

Gel retardation has provided the investigator with a tool that allows the separation and characterization of sequence-specific transcription factors from the myriad of cellular proteins (*see* Chapter 15). Using this technique in conjunc-

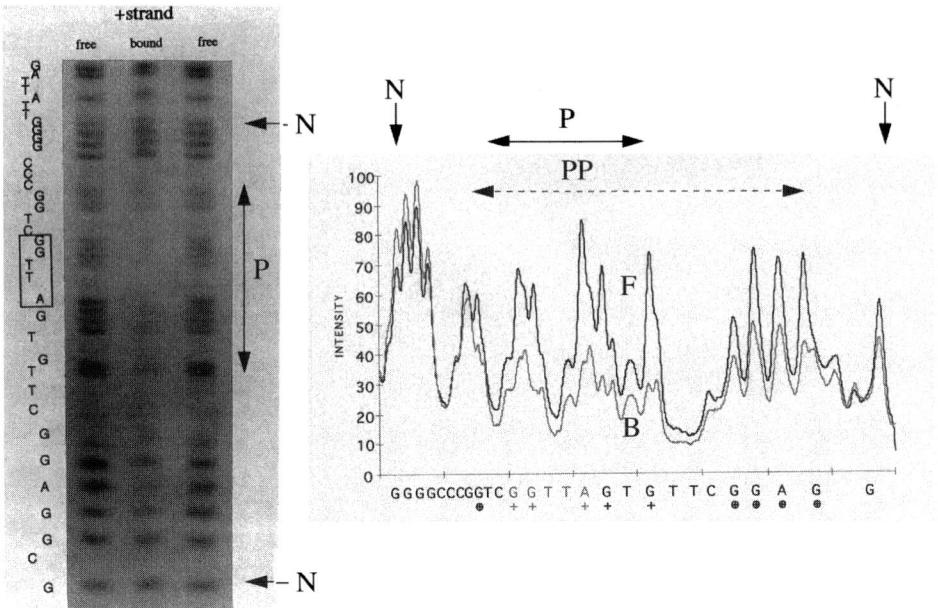

Fig. 1. A standard protection footprinting gel and subsequent data analysis, for one strand of a defined CCAAT-box, which is known to bind a transcription factor (3). Position N refers to the points of normalization where the bound (B) intensity is adjusted to that of the free (F). P refers to the protected region and PP that of the partially protected bases. The analysis of footprinting data by phosphorimaging provides a far clearer picture than conventional autoradiography.

tion with DNA footprinting, it has become possible to characterize the protein–DNA interactions of unpurified transcription factors. Provided that the binding constant for a particular protein–DNA interaction is sufficiently high to perform gel retardation, it is likely that DNA footprinting can be performed on the retarded complex. If the cis-acting DNA sequence is known, then the three techniques described below should give extensive information about the DNA interactions of the trans-acting factor.

The gel retardation based approach to protection footprinting is straightforward. The protein–DNA complex is modified or cleaved, prior to or after entering a band-shift gel. DNA is extracted from the gel by blotting and the band(s) of interest identified by autoradiography. Bands representing free and bound DNA are eluted, cleaved (if required) and separated using a denaturing sequencing gel (*see* **Fig. 2**) *(5)*. Protection assays demonstrate which sequences, or areas of DNA, are inaccessible to cleavage or modification resulting from protein coverage.

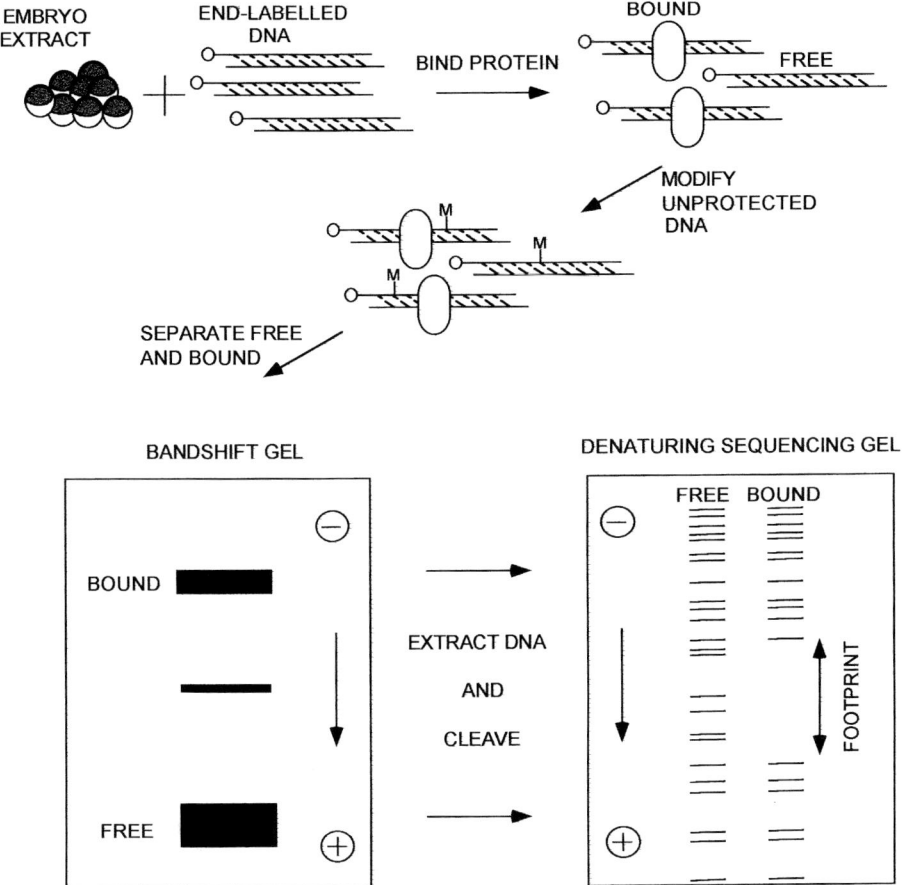

Fig. 2. Methylation protection overview. A radioactively end-labeled (plus or minus strand) DNA fragment containing a putative transcription factor binding site is incubated with crude embryo extract. Unprotected bases are modified immediately prior to gel retardation. Following electrophoresis and blotting "free" and "bound" DNA is recovered and cleaved. Fragments are separated using a high-percentage, denaturing sequencing gel and visualized using autoradiography. Bases protected from modification will not be cleaved; hence, the corresponding DNA fragment will be absent from the bound lane.

Interference footprinting demonstrates which bases or chemical groups interact with, or are in close proximity to, the transcription factor. DNA is modified prior to protein complex formation and separation by gel retardation. As with protection footprinting, the assumption is made that only one base is

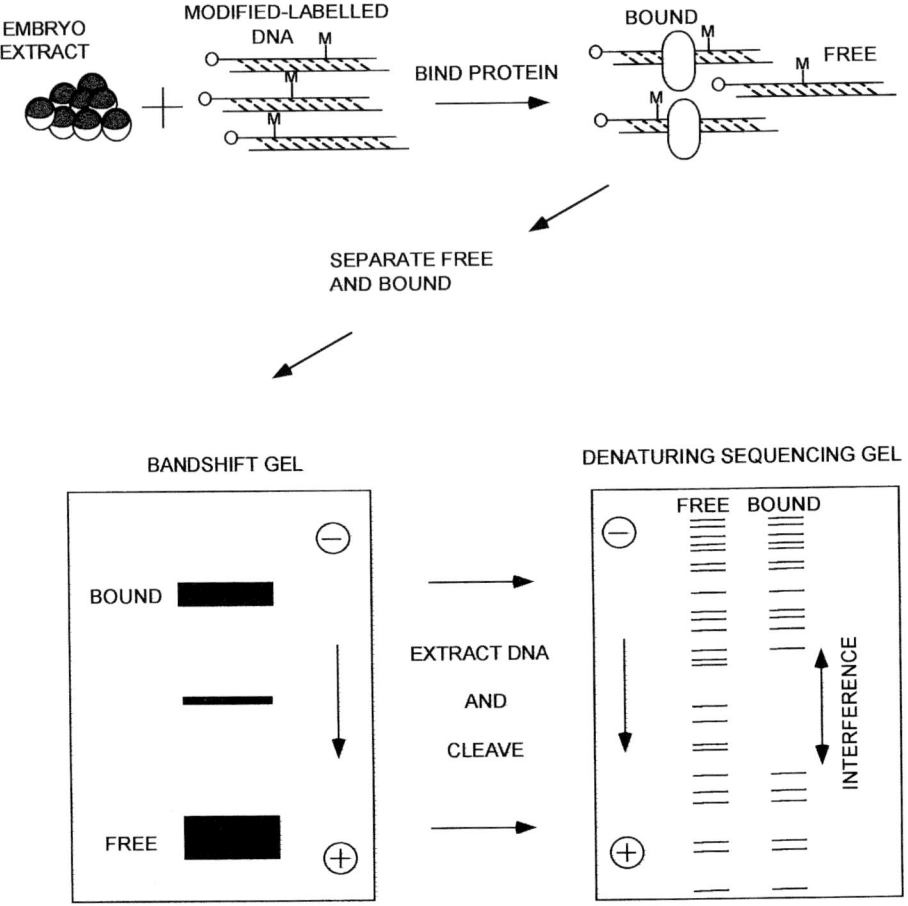

Fig 3. Methylation interference overview. A DNA probe which has been radioactively end-labeled on a defined stand is modified at a frequency of one base per fragment. This probe DNA is then incubated with crude embryo extract, bound DNA is separated by gel retardation. Following electrophoresis and blotting, DNA is recovered, cleaved and separated using a high-percentage, denaturing sequencing gel. DNA modified at sites that inhibit binding will not be present in the bound lane.

modified per DNA molecule. Species of DNA molecules modified such that they disrupt protein binding are found in the "unbound" DNA band. Likewise DNA modified at a noncritical site will be apparent in the "bound" band (*see* **Fig. 3**). The combination of protection and interference techniques allows the investigator to elucidate the "where and how" of transcription factor binding.

The first protocol described is that of the chemical nuclease copper orthophenanthroline (OP–Cu). Cleavage occurs at the phosphodiester backbone of nucleic acids at physiological pHs and temperatures *(6)*. Protection from $(OP)_2CU^+$-mediated cleavage implies that at least a portion of the protein occupies the minor groove (*see* **Note 1**). This OP–Cu footprinting is novel, as the scission reaction can take place within the polyacrylamide matrix following gel retardation. The second protocol is that of dimethyl sulfate (DMS) protection or methylation footprinting. This technique requires that the protein–DNA complex is briefly treated with DMS immediately prior to gel retardation. The manner of this chemical modification arises from a mild exposure of the DNA to the oxidizing agent DMS. Principally, methylation occurs on exposed guanine residues at the N7 position (major groove) and, to a lesser extent, adenine residues at the N3 position (minor groove). Methylated DNA is later cleaved by piperidine treatment. Although this technique does not provide the same resolution as the above *in situ* approach, DMS demonstrates a reduced sensitivity to conformational variability (*see* **Note 1**). The final technique is that of methylation interference using dimethyl sulfate. This assay identifies the extent of protein contacts to specified bases within the recognition sequence. Methylation interference is the most widely used of the footprinting techniques and is often used for comparison with other characterized factors (*see* **Note 2**) *(7)*.

2. Materials
2.1. Maxam and Gilbert Sequencing Reagents

1. DNase- and RNase-free water (Sigma, Poole, UK).
2. 90% Formic acid (Merck, Lutterworth, UK).
3. 10 M Piperidine (Merck).
4. 10 mg/mL Salmon sperm DNA (Boehringer Mannheim, Lewes, UK).
5. Parafilm.
6. Water-saturated butan-1-ol (Merck).
7. Heating block.
8. Variable-speed bench-top centrifuge.
9. 3 M Sodium acetate (pH 4.5–5.5) (Merck).
10. 100% and 70% Molecular-biology-grade ethanol (EtOH) (Sigma).

2.2. Copper Phenanthroline Reagents

1. 40 mM 1,10 Phenanthroline (0.079 g dissolved in 10 mL EtOH, prepared fresh).
2. 9 mM Cupric sulphate (0.022 g dissolved in 10 mL H_2O, prepared fresh).
3. 28 mM 2,9-Dimethyl-1,10-phenanthroline (0.104 g dissolved in 10 mL EtOH, prepared fresh).
4. 50 mM Tris–HCl, pH 8.0 (Sigma).
5. β-Mercaptoprionic acid (Sigma).

2.3. DNA Extraction and Preparation.

1. 3MM Paper (Whatman, Maidstone, UK).
2. DE-81 Paper (Whatman).
3. Western-blotting tank (Bio-Rad, Hemel Hemstead, UK).
4. 3 L 1X Tris-borate EDTA (TBE) Ultrapure: 0.089 M Tris-borate pH 8.3, 2 mM EDTA (National Diagnostics, Hull, UK).
5. Elution buffer: 1 M NaCl, 20 mM Tris pH 8, 1 mM EDTA.
6. Biophenol–chloroform–iso-amyl-alcohol (Camlab).
7. Chloroform (Merck).
8. 1 µg/µL tRNA (Boehringer Mannheim).
9. Spin-X tubes (Sigma).

2.4. Footprinting Gel Materials

1. Sequencing gel plates.
2. Sequencing gel tank. Acrylamide sequencing grade 40% (29:1) acrylamide:bisacrylamide solution (gas stabilized) (National Diagnostics).
3. Silane (Sigma).
4. Yellow sealing tape.
6. 10% (w/v) Ammonium persulfate (APS) (store solution at 4°C and use within 1 wk) (Sigma).
7. $N,N,N'N'$-Tetramethylethylenediamine (TEMED) (Sigma).
8. Urea (Sigma).
9. Formamide loading dye: 10 mL formamide, 10 mg xylene cyanol, 10 mg bromophenol blue, 200 µL 0.5 M EDTA pH 8.

2.5. Dimethyl Sulfate Footprinting Reagents

1. 10M Dimethyl sulfate (Sigma).
2. 1X TE (10 mM Tris pH 7.5, 1 mM EDTA).
3. Dimethyl-sulfate (DMS) buffer: 60 mM NaCl, 10 mM Tris-HCl, pH 8.0, 10 mM MgCl$_2$, 1 mM EDTA.
4. 250 mM DTT stored at –20°C (Sigma).

3. Methods
3.1. Preparation of G + A Ladder: Maxam and Gilbert

The preparation of a suitable marker is essential to orient and analyze footprinting gels. Both strands of the DNA fragment to be used in the footprinting assays should be sequenced. This protocol involves extensive use of radioactivity, always wear gloves, glasses, work behind protective screens and monitor for contamination.

1. Take 15–20 fmol of 5' end-labeled dsDNA in Eppendorf tubes (+/– labeled strands separately) (*see* Chapter 15 and **Notes 2** and **3**), to each tube add 1.5 µL 10 mg/mL Salmon sperm DNA and 10 µL Sigma water.

2. Chill the reaction on ice and add 1.5 µL 88% formic acid.
3. Incubate at 37°C for 14 min.
4. Chill on ice and add 1.0 M aqueous piperidine.
5. Secure the lid with parafilm (to avoid opening).
6. Heat to 90°C for 30 min; cover the tubes with a lead block.
7. Chill on ice and centrifuge for a few seconds.
8. Add 1 mL of water-saturated butan-1-ol and vortex vigorously.
9. Centrifuge at room temperature for 2 min at 14,000g.
10. Add 0.1 volume 3 M sodium acetate (pH 4.5–5.5) and vortex.
11. Add two volumes of ice cold 100% ethanol and invert.
12. Centrifuge for 20 min at 14,000g.
13. Carefully decant the ethanol, do not disturb the pellet.
14. Add 70% ethanol to the pellet and spin at 14,000g for 5 min.
15. Carefully decant the supernatant and dry the pellet.
16. Finally, Cerenkov count the pellets to calculate resuspension volume.
17. After resuspension, place the samples into new tubes and recount (*see* **Note 5**).

3.2. Copper Orthophenanthroline Protection Footprinting
3.2.1. Preparative Gel Retardation

Reference must be made to Chapter 15 for further information about performing gel retardation assays. Essentially, this protocol describes how to scale up a standard band shift in order to obtain higher levels of retarded complex.

1. Prepare and prerun two standard band-shift gels with the inclusion of an eight space (20 mm wide) preparative comb.
2. Prepare extraction buffer and chill on ice.
3. To two tubes of 50 embryos, add 450 µL extraction buffer and homogenize with a 1-mL pipet tip.
4. Add 500 µL 1,1,1-Trichlorofloroethane (Freon) to each tube, vortex, and centrifuge at 14,000g for 3 min.
5. Decant the upper phase into a fresh tube and store on ice.
6. Prepare two tubes of reaction mix as follows: Add 20 µL +/– labeled 40 fmol/µL probe, 20 µL 1 µg/µL poly(dI-dC)–poly(dI-dC), and 160 µL electrophoretic mobility shift assay (EMSA) reaction buffer.
7. Add 400 µL crude embryo extract to each tube and mix by inverting.
8. Incubate at 4°C for 15 min to allow binding.
9. Load eight lanes of 70 µL for both +/– probe reactions.
10. Run the gel at 200 V for 95 min (depending on probe length and composition, to retain free probe on the band-shift gel).

3.2.2. Cleavage Reaction

1. Prepare the reagents detailed in the materials **Subheading 2.2.** during the electrophoresis above.

2. Following gel retardation, remove the large gel plate and cut the bottom left corner to facilitate orientation.
3. Carefully place the gel in a clean plastic sandwich box and add 100 mL 50 mM Tris, pH 8.0.
4. To 8 mL of double-distilled water add 1 mL 40 mM 1,10-phenanthroline and 1 mL 9 mM cupric sulfate.
5. Add this mixture to the plastic container and agitate gently for 5 min.
6. Using a fume hood, add 50 µL β-mercaptoproprionic acid to 9.95 mL of double-distilled water.
7. Add this solution to the plastic container, agitate and incubate at room temperature for 15 min.
8. To stop the reaction, add 10 mL 28 mM 2,9-dimethyl-1,10-phenanthroline, agitate, and leave for 15 min.
9. Decant off the solution and wash several times with distilled water.

3.2.3. Extraction and Preparation of DNA

1. Cut four gel-sized pieces of 3MM paper and one piece of DE-81 paper, marking the paper for orientation.
2. Presoak blotting apparatus and paper in 1X TBE.
3. Assemble the treated gel with two pieces of 3MM paper each side, with the DE-81 paper closest to the gel-side facing the anode.
4. Transfer the DNA from the gel to the DE-81 paper at 300 mA for 2 h.
5. Following blotting, remove the DE-81, cover with saran wrap and expose to autoradiography at 4°C for 2 h (*see* **Note 5**).
6. Develop the autoradiograph and cut out the bands of interest (free and bound DNA) using a scalpel.
7. Overlay the autoradiograph and DE-81 paper and cut out the required bands.
8. Transfer the paper to an eppendorf tube, add 0.4 mL elution buffer and vortex vigorously.
9. Elute the DNA overnight at 37°C.
10. Remove the paper from the elution buffer by centrifuging tube contents through a Spin-X filter tube at 14,000g for 30 s.
11. Add 0.4 mL phenol–chloroform–iso-amyl-alcohol, vortex, and centrifuge for 3 minutes at 14,000g.
12. Decant the aqueous supernatant into a fresh tube.
13. Add 0.4 mL chloroform, mix by inverting, and spin for 30 s at 14,000g.
14. Decant the aqueous supernatant into a fresh tube.
15. Add 10 µg tRNA before ethanol precipitation of the DNA, as in **steps 10–17** of **Subheading 3.1.**
16. After resuspension of the pellets, load equal counts of free and bound DNA onto a prewarmed denaturing sequencing gel (below).

3.2.4. Preparing and Running the Footprinting Gel

The percentage denaturing gel depends on the length DNA fragment to be used, the following protocol assumes the probe oligonucleotide is 40–60 basepairs (*see* **Note 3**).

1. Prepare a 17.5% denaturing sequencing gel as follows: Add 43.75 mL 40% (29:1) acrylamide:bisacrylamide solution, 10 mL 10X TBE, 45 g urea.
2. Microwave the solution for 20 s and make up to 100 mL with distilled water.
3. Allow the solution to cool to room temperature.
4. Clean sequencing plates thoroughly with water and ethanol.
5. Silane treat the smaller plate in a fume hood.
6. Align the gel spacers (0.4 mm) and tape the sequencing plates together carefully, hold together with bulldog clips.
7. To the precooled gel solution add 200 µL APS and 20 µL TEMED.
8. Mix the solution and pour into the sequencing plates using a 50-mL syringe.
9. Place an inverted shark-tooth comb (10-mm wide-well) in the gel and allow the gel to set for 2 h.
10. Prerun the gel at 65 W for 1 h.
11. Load the samples as sets of three the order free, bound, free to aid later analysis, run the Maxam and Gilbert sequence of the same strand along side.
12. For a 40-basepair probe, run the Bromophenol Blue front three-quarters of the distance down the gel.
13. Carefully remove the top plate from the gel and place a used autoradiograph film onto the gel, removing air bubbles by smoothing the film down.
14. Invert the plate so that the film is now on the bottom, very carefully pull the film so that the gel and film are removed as one (*see* **Note 8**). Cover the gel with Saran Wrap and expose to autoradiography or a phosphorimager plate (*see* **Notes 9** and **10**).

3.3. Dimethyl Sulfate Protection Footprinting

Care must be taken during this reaction, as it involves the use of 10 M DMS, which is an extremely volatile and toxic reagent. Waste solutions containing DMS should be disposed of in 5 M NaOH.

1. A band-shift reaction should be performed as in **Subheading 3.2.1.**, but without the addition of EMSA buffer.
2. After the binding period, add 1 µL 10 M DMS.
3. Vortex briefly and leave at room temperature for 90 s.
4. Stop the modification by the addition of 0.1 volume 250 mM DTT and vortex briefly.
5. Quickly add 160 µL EMSA buffer and immediately load onto the band-shift gel (*see* **Note 6**).
6. Extract and prepare the DNA as in **Subheading 3.2.3., steps 1–15**. Resuspend the pellet in 10 µL 1X TE and cleave the modified bases with piperidine as in **Subheading 3.1.** excluding **steps 2** and **3**.
7. Run footprinting reactions on a denaturing sequencing gel as in **Subheading 3.2.4.**

3.4 Methylation Interference Footprinting

Care should be taken during this protocol, as samples are radioactive and contain 10M DMS. Work in a fume hood, with the necessary protection. Treat waste materials carefully; samples containing DMS and γ-^{32}P should be neutralized with 5 M NaOH and allowed to decay prior to disposal.

1. Prepare 100 μL 5' end-labeled 40-fmol/μL +/– strand DNA probes essentially as in **steps 4–8**, **Subheading 3.1.1.** of Chapter 15, using stock oligonucleotides of 2 pmol/μL. Remember to only label one stand per reaction.
2. Anneal each labeled strand to 2 μL 2 pmol/μL cold opposite strand by heating to 95°C for a few minutes and allow to cool overnight.
3. Each tube should now contain 102 μL of a +/– 5' end-labeled double-stranded DNA probe at 40 fmol/μL.
4. Add 4 μL sonicated salmon sperm DNA (1 μg/μL) to each tube and vortex briefly.
5. Ethanol precipitate each reaction as in **steps 10–15** of **Subheading 3.1.**
6. Resuspend the pellet in 5 μL 1X TE and chill on ice.
7. Add 190 μL chilled DMS buffer and vortex briefly.
8. Add 5 μL 10% DMS, vortex, and incubate at 20°C for 5 min.
9. Stop the modification reaction by adding 50 μL 250 m*M* DTT, mix, and place on ice.
10. Ethanol precipitate the modified DNA as above.
11. Carefully resuspend the DNA in 20 μL TNE buffer.
12. Using the modified DNA, perform a standard preparative band shift, as in **Subheading 3.2.1.**
13. Extract and prepare the DNA as in **Subheading 3.2.3.**, **steps 1–15**.
14. Resuspend the pellet in 10 μL 1X TE and cleave the modified bases as in **Subheading 3.1.** excluding **steps 2** and **3**.
15. Run footprinting reactions on a denaturing sequencing gel as in **Subheading 3.2.4.**

4. Notes

1. Cu–OP footprinting works on the principle that the retarded complexes are active, homogeneous and that the conformational state of the complex is not governed by buffer components, such as EDTA. Most scission events are observed from the tetrahedral cuprous complex cutting at the nucleoside C1 position in the minor groove. The efficiency of the Cu–OP is dependent on the secondary structure of the DNA. An apparent increased scission reactivity for a certain base may be due to a conformational variation at this position, allowing for greater accessibility of the tetrahedral cuprous complex to this site. Approximately 80% of this scission sequence-dependent reactivity has been attributed to the flanking nucleotide 5' and 3' of the cut site *(4)*.
2. Interference footprinting is thought of as the reverse of methylation protection footprinting, as the DNA is exposed to the modifying reagent prior to protein binding. The assumption is made that the amount of free DNA in the binding reaction vastly exceeds that of any retarded bands. The unbound fraction therefore contains every species of methylated DNA, whereas the retarded band contains species of methylated DNA that do not interfere with binding.
3. The use of synthetic oligonucleotides is strongly advised when performing footprinting assays coupled with gel retardation of crude extracts. The benefits of

using oligonucleotides over restriction fragments are numerous and include the avoidance of difficult subcloning and radiolabeling procedures, increased band-shift sensitivity and increased yields of labeled footprinted fragments. Oligonucleotides containing a centralized transcription factor binding site and flanking regions of 20 basepairs are adequate for band-shift-coupled footprinting.

4. It is strongly recommended that the integrity of synthesized DNA oligonucleotides are investigated prior to footprinting. This is achieved by end-labeling the plus and minus strand DNA separately and running them on a denaturing sequencing gel. If the lanes contain a high percentage of species that are not full length, gel purification of oligonucleotides is required. Gel purification of DNA fragments is achieved easily by UV shadowing.

5. It is necessary to Cerenkov count the precipitated footprinting reaction twice, as a proportion of the labeled DNA always remains bound to the Eppendorf tube. After the second Cerenkov count, it is possible to calculate loading volumes to give equal counts of free and bound DNA. It is suggested that similar amounts of the opposite strand reaction are loaded, only one autoradiograph exposure is then required to visualize both.

6. To avoid confusion, the positively charged DE-81 should be orientated prior to transfer. Mark which side is next to the gel and which way is up (in pencil). Also mark the 3MM paper positive or negative depending on which electrode it is next to. After the transfer of DNA, ensure that the autoradiograph film is orientated about the DE-81 paper to facilitate the cutting of bands later. This is most easily achieved by the use of fluorescent strips, although other techniques can be used.

7. If treatment of the protein–DNA complex leads to the ablation of retarded band, then it is likely that amino acids within the DNA binding domain of the transcription factor have been modified by the dimethyl sulfate, abolishing binding. This assumes that the transcription factor is either not bound prior to the DMS treatment or that the dissociation rate of the complex is relatively high.

8. Although fewer gels are lost, the use of wet gels have the disadvantage that only one or two exposures can be obtained because the urea comes out of solution when warmed to room temperature. Exposing wet gels to autoradiography has the benefit that no fixing and drying is required, which is often the failing point of high-percentage, fragile polyacrylamide gels. These techniques involve analyzing very small quantities of protein, hence very small quantities of labeled DNA. Autoradiographs often require exposure for a few weeks, as yields of 1000 cpm per lane are not uncommon; one or two exposures are often the maximum obtainable.

9. Densitometry is fast becoming redundant, most investigators now use a phosphorimager to analyze data. Although expensive, phosphorimaging has the following advantages: (1) The rate of exposure is 10 times faster than conventional X-ray film; (2) linearity of signal is five orders of magnitude, much greater than that of X-ray film; (3) data analysis is more user friendly.

10. Normalization of "free" and "bound" lanes are required for the analysis of footprinting data. A rule of thumb is that sequences 10 bases away from either

side of the protected region should be used for normalization. Simply calculate the percentage difference for each flanking region and normalize the lanes accordingly with respect to each other, the flanking regions should now be equal. This technique highlights true protected areas and hypersensitive regions of DNA. Areas closer than 10 bases to the protected region often demonstrate hypersensitivity because of relaxation or flexibility of the DNA from the bound transcription factor.

References

1. Galas, J. G. and Schmitz, A. (1978) DNase I footprinting: a simple method for the detection of protein–DNA binding specificity. *Nucleic Acid Res.* **5,** 3157.
2. Maxam, A. and Gilbert, W. (1980) Sequencing end-labeled DNA with base-specific chemical cleavages. *Methods Enzymol.* **65,** 499–560.
3. Sigman, D. S. and Chen, C. B. (1990) Chemical nucleases: new reagents in molecular biology. *Ann. Rev. Biochem.* **59,** 207–236.
4. Shaw, P. E. and Stewart, A. F. (1994) DNA-protein interactions, in *Methods in Molecular Biology*, vol. 30 (Kneale, G. G., ed.), Humana, Totowa, NJ, pp. 79–95.
5. Sigman, D. S., Graham, D. R., D'Aurora, V., and Stern, A. M. (1979) Oxygen-dependent cleavage of DNA by the 1,10-phenanthroline–cuprous complex. *J. Biol. Chem.* **254,** 12,269–12,272.
6. Dorn, A., Bollekens, J., Staub, A., Benoist, C., and Mathis, D. (1987) A multiplicity of CCAAT box binding proteins. *Cell* **50,** 863–872.
7. Orford, R. L., Robinson, C., Haydon, J., Patient, R. K., and Guille, M. J. (1998) The maternal CCAAT box transcription factor which controls GATA-2 expression is novel, developmentally regulated and contains a dsRNA-binding subunit. *Mol. Cell. Biol.* **18,** 5557–5566.

17

Mapping Protein–DNA Interactions Using In Vivo Footprinting

David Warshawsky and Leo Miller

1. Introduction

In vivo footprinting is a technique that enables one to detect protein–DNA interactions as they are occurring in a cell. The principle behind this technique is similar to the principle behind the in vitro footprinting technique: Both rely on the fact that a bound protein often causes its binding site to be protected from cleavage by an endonuclease or from modification by a chemical agent. The major difference between in vivo and in vitro footprinting is that in the first method, cleavage of the DNA molecule is carried out within the nucleus following the in vivo binding of proteins to DNA in the context of chromatin, whereas in the second method, cleavage occurs in the test tube with purified DNA and protein extracts (or purified proteins) that are incubated together. Furthermore, footprints and hypersensitive sites due to deformations of DNA in chromatin are also detected by in vivo footprinting. Hence, compared with data obtained by in vitro footprinting, data obtained by in vivo footprinting may be more significant and representative of the true events that occur in the living cell.

Various agents can be used as probes to detect protein–DNA interactions, including chemicals such as dimethyl sulfate (DMS) and nucleases such as deoxyribonuclease I (DNase I) and micrococcal nuclease (MNase). In this review, we describe the use of DNase I and MNase for in vivo footprinting, in conjunction with ligation-mediated polymerase chain reaction (LMPCR) used to reveal and to amplify the enzymatic DNA nicking pattern. LMPCR in vivo footprinting was first described by Pfiefer et al. *(1)* and Mueller and Wold *(2)*. Later, it was improved by the addition of magnetic extension product capture (EPC) *(3)*. The in vivo footprinting method described here is a modification of the above protocols and has worked well using small quantities of *Xenopus* genomic DNA *(4,5)*. Typically, 1–2 µg of DNA is used, which corresponds to 3×10^5 *Xenopus* cells; however, successful footprints were obtained with less then 100 ng of material. We have used this protocol to identify in vivo footprints at the promoter of the *Xenopus* 63-kDa keratin gene in epidermal cells as early as stage 44 (4-d-old tadpoles).

In the initial step, epidermal cells (which have been permeabilized) are treated with a low concentration of DNase I or MNase for several min. The low concentration and short treatment is used to cleave the DNA every 500 basepairs or so, generating a collection of fragments of different lengths. DNase I and MNase detect different types of DNA–protein interactions and therefore complement each other. DNase I binds to the minor groove of DNA and nicks the phosphate backbone *(6,7)*. It can detect protected bases within the minor groove. DNase I also nicks bases at DNA helix deformations such as bends and regions with an abnormal minor groove width *(7,8)*. Such irregularities may be caused by binding of proteins or result from interactions between proteins bound at different cis elements *(9,10)*. MNase binds and cleaves the phosphodiester backbone of DNA *(9,11)*. It provides data on accessibility of the backbone and can detect proteins which contact it. Following nuclease digestion, DNA is isolated from the treated cells. This DNA will now carry the "signature" of the protein–DNA interactions, or "in vivo footprints," as sites which are protected from cleavage. As controls, purified DNA cleaved in the test tube as well as DNA from cells that do not express the gene under investigation are used. Sites that were bound by protein will appear as regions in which nicking is reduced; thus, bands that correspond to that region are weaker in comparison to the controls. To localize the position of nicking within a region of DNA, a ladder of guanine cleavage is generated and run side by side with the experimental and control lanes (**Fig. 1**).

Conventional PCR cannot be used to amplify nicked DNA fragments because it requires that each molecule to be amplified have two ends with known sequences. Therefore, the signature of footprints at any given region of DNA is amplified using LMPCR in conjunction with EPC (**Fig. 2**). In the

---

Fig. 1. *(opposite page)* Epidermal specific in vivo DNase I footprints on the coding strand, in the region upstream of the transcription initiation site. Purified epidermal cells from the skin of stage 44 tadpoles, stage 62 tadpoles, and adult frogs, blood cells from adult frogs and XL177 epithelial cells *(4)* were permeabilized and treated with DNase I. After purification the DNA samples were subjected to LMPCR with magnetic extension product capture. Lane 1, protein-free *Xenopus* genomic DNA was treated with DMS to produce a guanosine ladder (G). Lane 2, protein-free *Xenopus* genomic DNA was treated in vitro with DNase I (F). Lane 3, XL177 epithelial cells (X). Lanes

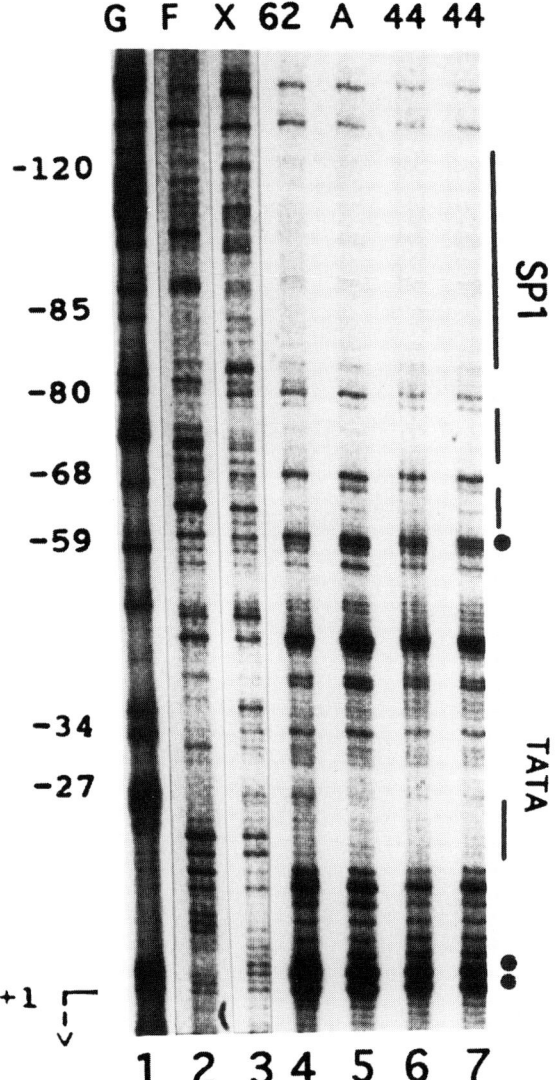

4–7, epidermal cells from stage-62 tadpoles (62), adult frogs (A), or stage-44 tadpoles. The primers used to obtain the patterns shown are given in Warshawsky and Miller *(4)*. The labeled products were separated by electrophoresis through a sequencing gel, dried and exposed to X-ray film. The DNA source is indicated on the upper part of the figure. Numbers indicate nucleotide positions relative to the 63-kDa keratin gene transcription initiation site (arrow). The position of the putative TATA box (–27 to –34 bp) and the SP1 site (–92 to –101 bp) are indicated by the TATA and SP1 labels. Vertical bars indicate maximal regions of protection from cleavage in epidermal cells. Solid circles indicate the positions of nucleotides which are hypersensitive to nicking.

Detection of single strand nicks by LMPCR

denature DNA, anneal biotinylated primer 1 and extend with Vent (exo-) to create blunt end

primer 1 (biotinylated)

desalt DNA with microcon 10 (*phosphorylate MNase treated DNA) ligate common linker to blunt end

incubate DNA with streptavidin coated magnetic beads, biotinylated molecules bind to beads

wash beads and elute non-biotinylated single stranded DNA with NaOH

Perform PCR with "nested" primer 2 and linker primer

primer 2

linker primer

remove 10% of the PCR products to new tube, perform 6 cycles with end labeled primer 3

primer 3 P 32

linker primer

LMPCR–EPC procedure, nicked DNA is denatured and a biotinylated gene-specific primer (primer 1) is annealed and extended with Vent (exo-) DNA polymerase, creating a blunt end at the nicking site. The region of interest is defined by this gene-specific primer. The primer-extended molecules are desalted and ligated to a synthetic, double-stranded common linker. Because MNase treatment creates DNA fragments without 5' phosphates, MNase-nicked DNA needs to be phosphorylated in vitro using T4 polynucleotide kinase following the primer-extension step. This linker is designed such that the duplex between the long and short oligomers is stable at ligation conditions but not at PCR temperatures, and the longer oligomer should have a comparable melting temperature to that of the second gene-specific primer. The DNA is incubated with streptavidin-coated magnetic beads. Only the primer-extended, biotinylated DNA will bind to the beads. The bulk of the DNA is removed by washing the beads, and the specifically bound DNA is eluted. The selection for extended molecules significantly reduces amplification of nonspecific DNA fragments in the following PCR. The eluted molecules represent all the fragments created by the in vivo cleavage of the genomic DNA. The eluted molecules are exponentially amplified by PCR with a second gene-specific primer (primer 2), located slightly downstream of primer 1 and the large oligomer of the common linker. Following PCR amplification, the fragments are indirectly end-labeled by extension of a third gene-specific primer that overlaps primer 2 and extends 3' to it, using several PCR cycles. The orientation of primers 1, 2, and 3 is discussed below (*see* **Note 1**).

2. Materials
2.1. General Reagents and Equipment
1. Phenol–chloroform–isoamyl alcohol (25:24:1).
2. Chloroform–isoamyl alcohol (24:1).
3. 3 *M* Na Acetate, pH 5.2.
4. TE buffer: 10 m*M* Tris–HCl, pH 7.4, 1 m*M* EDTA.
5. RNase A (Sigma).
6. Sequencing gel loading buffer: 90% formamide, 20 m*M* EDTA, 0.05% bromophenol blue, 0.05% xylene cyanol.
7. A 0.4-mm thick 6% Long Ranger sequencing gel (*see* **Note 7**).
8. Phosphate-buffered saline.

Fig. 2. *(previous page)* Schematic outline of ligation-mediated PCR in conjunction with magnetic extension product capture. For simplicity, only one DNA fragment is shown. Primer 1 is 5'-biotinylated. Primer 2 is "nested" to primer 1. Primer 3 is end-labeled ^{32}P. The second PCR primer is the long oligomer of the common linker. B = biotin; S = streptavidin.

2.2. In Vivo Nuclease Treatment

1. Nuclease buffer: 300 mM sucrose, 15 mM Tris–HCl, pH 7.5, 50 mM KCl, 15 mM NaCl, 5 mM MgCl$_2$, 1 mM CaCl$_2$, 0.5 mM 2-mercaptoethanol.
2. DNase I (Worthington): resuspend at 6 units/µL according to the directions of the manufacturer.
3. MNase (Worthington): resuspend at 10 units/µL according to the directions of the manufacturer.
4. 1% NP-40.
5. Stop solution: 2.5% sodium dodecyl sulfate (SDS), 50 mM EDTA, 100 mM Tris–HCl, pH 8, 50 mM NaCl, 2 mg/mL proteinase K, prepared fresh for each experiment.

2.3. In Vitro Nuclease Treatment

1. DNase I buffer: 40 mM Tris–HCl pH 7.7, 10 mM NaCl, 5 mM MgCl$_2$.
2. MNase buffer: 60 mM KCl, 15 mM NaCl, 15 mM Tris–HCl pH 7.5, 5 mM MgCl$_2$, 1 mM CaCl$_2$.
3. 0.5 M EDTA.
4. 100 mM EGTA.

2.4. In Vitro Dimethyl Sulfate–Piperidine Treatment

1. Lid-locks for 1.5-mL Eppendorf tubes.
2. DMS (Aldrich) (*see* **Note 8**).
3. DMS stop solution: 1.5 M Na acetate pH 7, 1 M β-mercaptoethanol, 100 µg/mL yeast tRNA. (Mix the Na acetate pH 7 and the β-mercaptoethanol and filter sterilize. Then add yeast tRNA to the final concentration.)
4. 1 M piperidine (Aldrich) (*see* **Note 8**).

2.5. Ligation Mediated PCR

1. BSA (3 mg/mL).
2. dNTP mix: dATP, dCTP, dGTP, and dTTP, each at 2.5 mM.
3. Vent (*exo-*) DNA polymerase (New England Biolabs).
4. 10X Vent (*exo-*) buffer (New England Biolabs).
5. Gene-specific primers 1, 2, and 3 (*see* **Notes 1** and **4**).
6. Primer-extension master mix (per reaction): 4 µL 10X Vent (exo-) buffer, 1 µL gene specific Primer 1 (0.3 pmol/µL), 1.3 µL BSA (3 mg/mL), 4 µL 2.5 mM dNTP mix, H$_2$O to 18 µL, 2 µL Vent (*exo-*) DNA polymerase (2 units/µL).
7. Spin columns: Microcon 10 (Amicon).
8. T4 Polynucleotide kinase.
9. 10X Kinase buffer (supplied with enzyme by manufacturer).
10. 10 mM ATP.
11. T4 DNA ligase (New England Biolabs).
12. 10X Ligase buffer (New England Biolabs).
13. Taq DNA polymerase.
14. Taq polymerase 10X buffer.

15. Common linker (*see* **Note 5**).
16. PCR gems (Perkin Elmer).
17. Lower Phase" mix (per reaction): 2 µL 10X PCR buffer, 2 µL 25 m*M* MgCl$_2$, 2 µL 5 m*M* dNTPS, 1 µL 10 pmol/µL gene-specific primer 2, 1 µL 10 pmol/L linker primer and 12 µL H$_2$O.
18. End-labeling mix (15 µL per reaction): 1.5 µL 10X buffer, 1.2 µL 25 m*M* MgCl$_2$, 0.75 µL 5 m*M* dNTP, 2 µL end-labeled primer 3 (1 pmol/µL) 1 unit Taq and H$_2$O to 15 µL.
19. End-labeling stop solution (85 µL per reaction): 10 µL 3 *M* Na Acetate pH 5.2, 2 µL 0.5 *M* EDTA, 1 µL tRNA, 5 µg/µL and 72 µL H$_2$O.

2.6. Extension Product Capture

1. Streptavidin-coated magnetic beads (Dynal).
2. Magnetic separation stand (Dynal).
3. 2X Binding buffer: 6 *M* NaCl, 10 m*M* Tris–HCl pH 7.7 and 1 m*M* EDTA.
4. Washing buffer: 2 *M* NaCl, 10 m*M* Tris–HCl pH 7.7, and 1 m*M* EDTA.
5. 1 *M* Tris–HCl, pH 7.7.
6. 5 mg/mL Yeast tRNA.

3. Methods

3.1. Summary of Steps in the In Vivo Footprinting Technique

1. Purified cells of interest are divided into aliquots, permeabilized, and treated with three different concentrations of DNase I or MNase.
2. Cells that do not express the studied gene are treated in the same manner in order to generate a negative control.
3. Protein-free genomic DNA is nicked with DNase I or with MNase to generate controls and with DMS–piperidine to generate a G-ladder of the region to be analyzed.
4. The cleaved DNA is harvested, subjected to LMPCR with EPC and separated on a sequencing gel.

3.2. Permeabilization of Cells and Enzymatic Cleavage of Genomic DNA

1. Collect 10^6 experimental cells and the same number of cells that do not express the gene (as a negative control). (It is possible to use fewer cells; *see* **Note 2**)
2. Wash the cells with 3 mL of ice-cold PBS by resuspending them in the buffer and spinning them down in a centrifuge. Repeat three times and resuspended the final cell pellet in 600 µL of ice-cold nuclease buffer, keep on ice.
3. DNase I cleavage: Prepare separate 1.5-mL Eppendorf tubes with 6, 20, and 60 units of DNase I in nuclease buffer in a final volume of 64 µL (a final concentration of 18, 60, or 180 U/mL in the reaction).
4. MNase cleavage: Prepare separate 1.5 mL Eppendorf tubes with 10, 25, and 83 units of MNase in nuclease buffer in a final volume of 64 µL (a final concentration of 30, 75, or 250 units/mL in the reaction).

5. Transfer 200 μL of the cell suspension to each of the tubes containing the nuclease.
6. Permeabilize the cells by adding 66 μL of 1% NP-40 (final concentration of 0.2%) and mix by pipetting. Incubate for 5 min at room temperature.
7. Stop the reaction by adding 80 μL of stop solution. Digest the cellular proteins overnight at 37°C.
8. Add one volume of phenol/chloroform/isoamyl alcohol, vortex, microcentrifuge the tube at 14,000g and remove the supernatant to a new tube.
9. Add 1 volume chloroform–isoamyl alcohol, vortex, microcentrifuge the tube at 14,000g and remove the supernatant to a new tube.
10. Add 0.1 volume 3 M NaOAc, pH 5.2 and 2.5 volumes of 100% ethanol, mix well, and place on crushed dry ice for 30 min. Microcentrifuge the tube at 14,000g for 20 min at 4°C.
11. Remove the supernatant, wash the pellet with 75% ethanol, and microcentrifuge for 10 min at room temperature. Remove the supernatant, air-dry the pellet for few minutes, and resuspend it overnight in 100 μL TE.
12. Digest the RNA with 50 mg/mL RNase A (Sigma) at 65°C for 20 min followed by an additional 30 min at 37°C.
13. Purify the DNA as in **steps 7–10**. Resuspend the DNA in 20 μL H$_2$O.

3.3. DNase I Digestion of Protein-Free Genomic DNA

1. Resuspend 60 μg of purified genomic DNA in 400 μL of DNase I buffer.
2. Aliquot 200 μL into two 1.5-mL tubes. Add DNase I to a concentration of 2 units/mL for the first tube and 5 units/mL for the second and incubate for 5 min at 23°C.
3. Stop the reaction by adding EDTA to 10 mM.
4. Add 250 μL of phenol–chloroform–isoamyl alcohol, vortex, microcentrifuge the tube at 14,000g for 5 min and remove the supernatant to a new tube.
5. Add 250 μL chloroform–isoamyl alcohol, vortex, microcentrifuge the tube at 14,000g for 5 min, and remove the supernatant to a new tube.
6. Add 0.1 volume 3 M NaOAc, pH 5.2, and 2.5 volumes of 100% ethanol and place on crushed dry ice for 30 min. Microcentrifuge the tube at 14,000g for 20 min at 4°C.
7. Remove the supernatant, wash the pellet with 75% ethanol and microcentrifuge for 10 min at room temperature. Remove the supernatant, air-dry the pellet for few minutes, and resuspend it overnight in TE (at a concentration of 2 μg/μL).

3.4. MNase Digestion of Protein-Free Genomic DNA

1. Resuspend 60 μg of purified genomic DNA in 400 μL of MNase buffer.
2. Aliquot 200 μL into two 1.5-mL tubes. Add MNase to a concentration of 20 units/mL for the first tube and 50 units/mL for the second, incubate for 5 min at 23°C.
3. Stop the reaction by adding EDTA to 10 mM and EGTA to 2.5 mM.
4. Add 250 μL of phenol–chloroform–isoamyl alcohol, vortex, microcentrifuge the tube at 14,000g for 5 min, and remove the supernatant to a new tube.

5. Add 250 µL chloroform/isoamyl alcohol, vortex, microcentrifuge the tube at 14,000g for 5 min, and remove the supernatant to a new tube.
6. Add 0.1 volume 3 M NaOAc, pH 5.2 and 2.5 volumes of 100% ethanol and place on crushed dry ice for 30 min. Microcentrifuge the tube at 14,000g for 20 min at 4 °C.
7. Remove the supernatant, wash the pellet with 75% ethanol, and microcentrifuge for 10 min at room temperature. Remove the supernatant, air-dry the pellet for few min on the bench, and resuspend the pellet overnight in TE (at a concentration of 2 µg/µL).

3.5. DMS Treatment of Genomic DNA

1. Place approximately 50 µg of *Xenopus* genomic DNA in a 1.5 mL tube. Bring the volume to 175 µL with TE.
2. Prepare ice-cold DMS stop solution, a bucket of dry ice, and 100% ethanol prechilled on dry ice.
3. Make a 1% DMS solution by adding 5 µL DMS to 495 µL H_2O (*see* **Note 8**).
4. Add 25 µL of 1% DMS to the genomic DNA. Incubate for 2 min at room temperature. Mix thoroughly by gentle vortexing for about 30 s.
5. Terminate the reaction by adding 50 µL of ice-cold DMS stop solution and immediately add 750 µL of prechilled 100% ethanol. Plunge the tube into crushed dry ice and leave it there for 30 min.
6. Microcentrifuge the tube at 14,000g for 20 min at 4°C, remove the supernatant, wash the pellet with 75% ethanol, and microcentrifuge at 14,000g for 10 min at room temperature.
7. Remove the supernatant but leave a few microliters of ethanol with the pellets. Do not let the pellet dry. Add 200 µL of piperidine to the pellets (*see* **Note 8**). Resuspend the DNA by incubating at room temperature with intermittent vortexing. Make sure that the pellet has been completely dissolved in the piperidine. An undissolved pellet will appear as a clear, floating lens in the piperidine.
8. After pellets are completely dissolved, centrifuge the tubes for 2 s, put a lid lock on the tubes, and incubate them at 90°C for 30 min. The hot piperidine cleaves the DNA at methylated guanines, denatures the DNA, and destroys contaminating RNA.
9. Resuspend the DNA pellets in 360 µL TE. Add 40 µL of 3 M NaOAc, pH 5.2 and 2.5 volumes of 100% ethanol and place on crushed dry ice for 30 min. Microcentrifuge the tube at 14,000g for 20 min at 4°C.
10. Remove the supernatant and repeat **step 9**.
11. Wash the pellet with 75% ethanol and microcentrifuge for 10 min at room temperature. Remove the supernatant, air-dry the pellet for a few minutes, and resuspend the pellet in 50 µL H_2O.
12. Dry in a Speedvac for about 1 h (until the pellet is completely dry) and resuspend the pellet in H_2O (at a concentration of 1 µg/µL).
13. Place the tubes in a microcentrifuge and spin at 14,000g for 10 min at room temperature. A gelatinous pellet may be present. Transfer the supernatant (containing the DNA) to a new tube.

3.6. Primer Extension

1. Prepare a sufficient amount of primer-extension master mix (number of reactions +1, 20 µL per reaction). Keep the mix on ice.
2. Denaturing step: place each DNA sample from **Subheading 3.2., step 13, Subheading 3.3., step 7, Subheading 3.4., step 7,** or **Subheading 3.5., step 13** in a clean 0.5-mL tube. Bring the volume of the DNA to 20 µL with H$_2$O (if needed) and add 20 µL of primer-extension master mix. Mix by pipetting and place the tubes in a thermocycler, preheated to 95°C, for 5 min (*see* **Note 3**).
3. Hybridization step: Incubate at primer specific temperature (*see* **Note 4**) for 30 min.
4. Extension step: Incubate at 76°C for 10 min and transfer to ice.
    Note: After this point, keep at low temperature at all times to prevent polymerase activity.
5. Desalting step: To the primer-extended DNA, add TE to 500 µL and spin at 4°C using a microconcentrator, such as microcon 10 (Amicon), until about 5–10 µL remain (about 40 min). Add 470 µL H$_2$O and repeat. Continue to spin until the final volume of the DNA is equal to or smaller than 19 µL. Collect the DNA into a new 1.5-mL tube.

3.7. Phosphorylation of MNase-Treated DNA

This step is not performed with DNase I or DMS treated DNA.

1. In a 1.5-mL tube, combine: 19 µL MNase treated genomic DNA (the primer-extended, MNase-treated DNA from **step 5** of **Subheading 3.6.**), 2.5 µL 10X kinase buffer, 0.25 µL 10 m*M* ATP, H$_2$O to 25 µL, and 1 µL T4 polynucleotide kinase. Incubate 1 h at 37°C.
2. Incubate at 75°C for 10 min to inactivate the kinase. The phosphorylated DNA is in a buffer that is compatible with ligase buffer.

3.8. Ligation

Keep cold at all times to prevent linker denaturation.

1. DNase I-treated DNA: To the desalted, primer-extended DNA, add 2 µL 10X ligase buffer, 3 µL (20 pM/µL) common linker (*see* **Note 5**), ice-cold H$_2$O to 19 µL and 1 µL ligase (6 Weiss units/µL).
2. MNase-treated DNA: To the phosphorylated, primer-extended DNA add 3.5 µL 10X ligase buffer, 5 µL (20 pM/µL) common linker, ice-cold H$_2$O to 49 µL, and 1 µL ligase (6 Weiss units/µL).
3. Incubate overnight at 17°C.
4. Spin the tube briefly in microcentrifuge and add 10 m*M* EDTA to a final volume of 75 µL.

3.9. Extension Product Capture with Magnetic Beads

1. Mix streptavidin coated magnetic beads gently by pipetting up and down and place 37.5 µL of beads (375 µg) in a 1.5-mL Eppendorf tube.

2. Place the tube in a magnetic separation stand, aspirate the supernatant and discard.
3. Wash the beads twice with 2 M NaCl, 10 mM Tris–HCl, pH 7.7, 1 mM EDTA. Resuspend the beads in 75 µL of 6M NaCl, 10 mM Tris–HCl pH 7.7, 1 mM EDTA.
4. To the washed beads, add 75 µL of DNA that has been ligated to the common linker and mix by pipetting gently. (Avoid losing beads that will stick to the pipette). Incubate for 20 min in a 48°C water bath (mix by pipetting every 5 min).
5. Place the tube in a magnetic separation stand, aspirate the supernatant, and discard.
6. Wash the beads twice with 200 µL of 2 M NaCl, 10 mM Tris–HCl pH 7.7, 1 mM EDTA.
7. To elute the nonbiotinylated DNA strand, add 50 µL 0.15 N NaOH to the washed beads and incubate for 12 min at 37°C.
8. Transfer the eluted DNA to a new tube containing 50 µL 0.15 M HCl and mix by pipetting.
9. Add 10 µL 1 M Tris–HCl pH 7.7, vortex, and spin down. Add 5 µg tRNA 13 µL 3 M NaOAc, vortex and spin down. Add 2.5 volumes of 100% ethanol and place on crushed dry ice for 30 min. Microcentrifuge the tube at 14,000g for 20 min at 4°C. Wash the pellet with 75% ethanol and microcentrifuge for 10 min at room temperature. Dry the pellet in a speedvac and resuspend in 20 µL H_2O.

3.10. PCR Amplification Using "Hot-Start" PCR with PCR Gems

It is vital to use hot-start PCR in order to minimize background from nonspecific amplification.

1. Prepare a sufficient amount of the "lower phase" master mix (number of reactions +1).
2. Prepare the "upper phase" for each reaction by mixing 3 µL 10X PCR buffer, 20 µL of the eluted DNA from **step 9** above, 8.75 µL H_2O, and 0.25 µL Taq polymerase (5 U/µL).
3. For each reaction place 20 µL of the "lower-phase" mix in a 0.5-mL tube.
4. Put a wax PCR gem on top of the lower phase, place in a thermal cycler, and heat to 80°C for 5 min, then cool to 4°C.
5. Add the upper phase on top of the hardened wax layer; place tube in thermal cycler.
6. Set the machine for an initial incubation at 95°C for 5 min followed by X cycles (*see* **Note 6**) with the following parameters:
    a. 95°C for 1 min.
    b. Annealing temperature for 1 min (*see* **Notes 1** and **4**).
    c. 75°C for 3 min, auto extension **step 2–5** s/cycle.
    d. A 10-min final extension at 75°C followed by a soak at 4°C.
7. Store at –20°C.

3.11. End-Labeling of LMPCR Products with ^{32}P End-Labeled Gene-Specific Primer 3

1. Prepare end-labeled primer 3 in a 1.5-mL tube containing
    2.5 µL (10 pmol/µL) of gene-specific primer 3
    2.5 µL 10X T4 polynucleotide kinase buffer

15 μL (10 μCi/μL) of fresh γ-^{32}P-ATP (150 μCi/25 pmol primer)
   20 units T4 polynucleotide kinase
   H$_2$O to 25 μL
   Incubate at 37°C for 60 min, then 75°C for 10 min to inactivate the kinase. Remove unincorporated radionucleotide with a commercial sephadex column. Resuspend the labeled primer at 1 pmol/μL.
2. Prepare a sufficient amount of end-labeling master mix (number of reactions +1).
3. Place 5 μL of PCR-amplified DNA from **step 7** of **Subheading 3.10.** into a new 0.5-mL tube.
4. Add 15 μL of end-labeling mix, mix by pipetting, add a drop of mineral oil, and spin down. Place the tube in a thermocycler preheated to 95°C and carry out seven rounds of PCR with the following parameters:
   a. 95°C for 1 min.
   b. Annealing temperature 1 minute (*see* **Notes 1** and **4**).
   c. 75°C 3 min, auto extension step 2-5 s/cycle.
   d. A 10-min final extension at 75°C followed by a soak at 4°C.
5. Prepare a sufficient amount of stop solution master mix (number of reactions +1) and add 85 μL to each tube from above.
6. Prepare 1.5-mL tubes with 150 μL phenol–chloroform–isoamyl alcohol.
7. Transfer the end-labeled DNA to the above tube, vortex, and centrifuge at 14,000*g* for 10 min at room temperature. Place the supernatant in a fresh tube, add 2.5 volumes of 100% ethanol, and place on crushed dry ice for 30 min. Microcentrifuge the tube at 14,000*g* for 20 min at 4°C. Wash the pellet with 75% ethanol and microcentrifuge for 10 min at room temperature.
8. Dry the pellet in a Speedvac and resuspend it in 10 μL loading buffer (90% formamide, 20 m*M* EDTA, 0.05% Bromophenol Blue, 0.05% xylene cyanol) by vortexing. Centrifuge briefly and check that the DNA is resuspended by removing the sample from the tube with a pipette and using a Geiger counter to check that the radioactivity is not left behind.
9. Prepare a 0.4-mm-thick 6% sequencing gel (*see* **Note 7**), denature the DNA for 5 min at 90°C, and load 2.5 μL from each reaction in adjacent lanes. Freeze the remaining aliquot at –70°C.

4. Notes

1. Primer design and orientation. Primer 1 should have an annealing temperature of 55°C or higher. Primer 2 should have an annealing temperature of 60°C or higher and primer 3 should have an annealing temperature of 2°C or more higher than primer 2. Primer 1 should be within about 200 nucleotides from the region under investigation. Primer 2 is located downstream of primer 1, and primer 3 should overlap primer 2 over at least half of the nucleotides of primer 2 and extend downstream of it by about 10 nucleotides.
2. The optimal number of cells per reaction is 3×10^5 (yielding about 2 μg of genomic DNA). However as little as 10,000 cells (yielding 100 ng of genomic DNA) can be used to detect footprints.

3. Primer extension with vent (*exo-*) DNA polymerase. Always use a thermocycler preheated to 95°C. After 5 min, working quickly, place the tubes in a microcentrifuge, spin for 2 s, and immediately place back in the cycler for annealing and extension.
4. Primer testing and optimization of annealing temperatures. For each primer pair, the approximate annealing temperature was first estimated using the formula $T_m = 4(G+C)+2(A+T)$. The optimal annealing temperature is determined empirically using a primer that is located 200–600 nucleotides downstream of the region of interest and has an annealing temperature of 68°C or higher, with plasmid and genomic DNA as templates. From our experience a magnesium concentration of 2 m*M* works well for all primer combinations. For primers 2 and 3, use Taq polymerase, for primer 1, use Vent (*exo-*) polymerase.
5. Common linker. The common linker is designed such that the duplex between the long and short oligomers is stable at ligation conditions but not at PCR temperatures, and the longer oligomer should have a comparable melting temperature to that of the second gene-specific primer. Linkers are prepared by annealing a 25-mer, 5' GCG GTG ACA CGT GAG ATC TGA ATT C 3', to an 11-mer, 5' GAATTCAGATC 3', at a concentration of 20 pmol/L each, by heating to 95°C for 3 min and gradually cooling to 4°C over a time period of 3 h. Linkers can be stored at –20°C for at least 1 yr. Keep on ice after thawing.
6. The number of cycles used depends on the amount of DNA used, the extent of DNA nicking and the reaction conditions. If the LMPCR procedure was carried out with 1–2 µg of DNA, 17–19 cycles are usually sufficient. Up to 26 cycles can be used if needed. The optimal number of cycles may have to be determined empirically.
7. We recommend the use of "Long Ranger" solution (T. J. Max) for preparation of the sequencing gel.
8. Handle DMS and piperidine with care. Use gloves and work in a fume hood.

References

1. Pfiefer, G. P., Stiegerwald, S. D., Mueller, P. R., Wold, B., and Riggs, A. D. (1989) Genomic sequencing and methylation analysis by ligation mediated PCR. *Science* **246,** 810–813.
2. Mueller, P. R. and Wold, B. (1989) In vivo footprinting of a muscle specific enhancer by ligation mediated PCR. *Science* **246,** 780–786.
3. Tormanen, V. T., Swidersky, P. M., Kaplan, B. E., Pfeifer, G. P., and Riggs, A. D. (1992) Extension product capture improves genomic sequencing and DNase I footprinting by ligation-mediated PCR. *Nucleic Acids Res.* **20,** 5487–5488.
4. Warshawsky D. and Miller L (1995) Tissue-specific in vivo protein–DNA interactions at the promoter region of the *Xenopus* 63 kDa keratin gene during metamorphosis. *Nucleic Acids Res.* **23,** 4502–4509.
5. Warshawsky, D. and Miller, L. (1997) In vivo footprints are found in the *Xenopus* 63 kDa keratin gene promoter prior to the appearance of mRNA. *Gene* **189,** 209–212.

6. Suck, D. and Oefner, C. (1986) Structure of DNase I at 2. 0A resolution suggests a mechanism for binding to and cutting DNA. *Nature* **321,** 620–625.
7. Hogan, M. E., Robertson, M. W., and Austin, R. H. (1989) DNA flexibility variation may dominate DNase I cleavage. *Proc. Natl. Acad. Sci. USA* **86,** 9273–9277.
8. Zhang, L. and Gralla, J. D. (1989) In situ nucleoprotein structure at the SV40 major late promoter: melted and wrapped DNA flank the start site. *Genes Dev.* **3,** 1814–1822.
9. Jackson, R. J. and Benyajati, C. (1992) *In vivo* stage- and tissue-specific DNA–protein interactions at the D. melanogaster alcohol dehydrogenase distal promoter and adult enhancer. *Nucleic Acids Res.* **20,** 5413–5422.
10. Van Der Vliet, P. C., and Verrijzer, C. P. (1993) Bending of DNA by transcription factors. *BioEssays* **15,** 25–32.
11. Drew, H. R. (1984) Structural specificities of five commonly used DNA nucleases. *J. Mol. Biol.* **176,** 535–557.

Index

A

activin, 2
adeno-associated virus, 168
alkaline phosphatase buffer, 59
alkaline phosphatase color reaction, 62
alkaline phosphate substrates, 57
anaesthetising embryos, 143
animal cap assay, 1
animal cap assay, experimental
    considerations, 8
animal cap excision, 6
animal cap sectioning, 12
annnealing temperature, 49
anti-digoxygenin, 62
anti-fluorescein, 62
antibody specificity, 95
antibody storage, 80, 96
antisense oligonucleotides, 112
atypical epidermis, 1
automated microinjection, 112

B

β-galactosidase, 113,135,162
β-galactosidase staining, 140, 146, 158, 163
backfilling needles, 127
band shift, 175
biotin-streptavidin, 83
blastocoel, 5
bleaching embryos, 61, 66, 91, 93
blebbing, 142
blockage of needles, 130
blocking reagent, 59

C

caging effect, 184
calcium-free Ringer's solution, 16
calcium-magnesium-free medium, 9
cardiac actin promoter, 137, 168
CAT, 135
cDNA libraries, 9
chordin, 25
chromatin structure, 111
cleavage of genomic DNA, 205
collagenase, 116
competitive PCR, 53
competitor templates, 53
confocal microscopy, 85
copper orthophenanthroline, 191
copper orthophenanthroline and
    secondary structure, 196
copper orthophenanthroline
    footprinting, 193
COS 7 cells, 23
cryostat, 81
cycle numbers, 49
cytomegalovirus promoter, 136, 169
cytoskeletal actin promoter, 136

D

DE-81 paper, 197
dechorionation, 131
deionized formamide, 71
dejellying, 3, 141, 170
Denhardt's reagent, 71
dexamethasone, 102
digoxygenin detection, 74
dissociated cells, 3, 9, 157, 161
DMS treatment, 207
DNA amplification, 125
DNA concentrations for injection, 131
DNA footprinting in vitro, 187
DNA footprinting in vivo, 199

DNA injection in Xenopus embryos, 133
DNA preparation, 121, 131
DNA template for transcription, 104
DNase I digestion, 206
DNase I specificity, 200
dominant interfering mutants, 99
dominant negative mutants, 112
dominant negative protease, 101
dominant negative receptors, 100
dominant negative signaling molecules, 100
dominant negative transcription factors, 101
donor yolk cells, 20
dorsal lymph sac, 117
double antibody labeling, 85, 94
Double WISH in *Xenopus,* 62
Double WISH in zebrafish, 65

E

ectopic expression, 135
EF-1α, 30, 55, 136
embedding wax, 71
embryo microinjection, 111
embryo preparation, 117
embryonic shield, 21
EMSA, 175, 188
EMSA gel, 181
EMSA probe, 183
engrailed transcription repressor domain, 101
epitope tagging, 99
explant fusion, 10
expression cloning, 111
expression screening, 9
extension product capture, 208
extraction of DNA from EMSA gels, 194

F

fast red, 67
fate mapping, 148
fate of injected DNA, 125

fate of injected DNA in *Xenopus*, 134, 137, 146
fate of injected RNA in *Xenopus*, 137
FGF receptor, 55
fibroblast growth factor, 2
fixatives, 86
fixing and storing *Xenopus* embryos, 61, 91, 93
fixing and storing zebrafish embryos, 63, 81
fixing and storing zebrafish larvae, 83
flag tag, 102
fluorescein-labeled probes, 66
fluorescent substrates, 57
footprinting analysis, 197
footprinting gel, 194
freeze substitution, 87

G

gain of function mutations, 99
GAPDH, 55
gel purification of oligonucleotides, 197
gel shift, 175
gelatin-albumin mix, 94
germinal vesicles, 120
glucocorticoid receptor, 102
goosecoid, 20, 21, 102
goosecoid promoter, 160
green fluorescent protein, 135, 164
Groucho, 101
guanadinium-isothiocyanate, 35, 43

H

half-caps, 3
hemagglutanin epitope, 102
histone H4, 50, 55
hooked glass needle, 18
hormone-binding domain, 102
human chorionic gonadothrophin, 115
hybridization, 39

I

immunohistochemistry, 77, 89, 93

Index

in situ hybridization, 72
in situ hybridization to sections, 69
in vitro transcription, 104
in vitro translation, 105
interference footprinting, 189
intracellular localisation, 111

L

labeling sectioned zebrafish, 84
labelling oligonucleotides, 180
larvae immunohistochemistry, 79
Leibovitz L-15 medium, 115, 161
LiCl precipitation, 32
ligation-mediated PCR, 199
lineage tracer, 99
linearizing DNA templates, 75
low-calcium medium, 7
luciferase, 135, 161

M

magnetic extension product capture, 199
maleic acid buffer, 59
Marc's modified Ringer's, 3, 140, 170
Max, 55
Maxam and Gilbert sequencing, 192
MBS, 114, 120
MEMFA, 58
mesoderm induction, 2, 19
methyl cellulose, 26
methylation interference, 190, 195
methylation protection, 189, 195
micrococcal nuclease digestion, 206
micrococcal nuclease specificity, 200
microinjection, 118, 129
microinjection chamber, 129
microinjection needles, 127
microinjection of zebrafish, 125
modified Barth's saline, 114
monoclonal antibodies, 79, 89
mosaic expression, 125, 157
MS222, 80, 120
multiplex PCR, 41, 45

Murray's clearing solution, 159
myc epitope, 102

N

Nanoject, 141
NASCO, 170
neural induction, 21
new Fuschin Red, 96
no tail, 20, 21
noggin, 25
nonimmune serum, 96
nuclear extracts, 178, 183

O

obtaining oocytes, 115
oligo (dT), 44
oocyte microinjection, 111
oocyte selection, 116
ornithine decarboxylase, 55
ovary removal, 116

P

PBSTw, 58
pCDM8, 25
PCR, 209
PCR mastermix, 46
PCR primer design, 48
PCR reaction, 45
pCS2+, 171
perinuclear, 169
permeabilization of cells, 205
permeabilization of larvae, 83
peroxidases, endogenous, 82
phosphate-buffered saline, 16
phosphor imaging, 47
phosphorylation of DNA, 208
poly(dI-dC), 178
poly-L-lysine coated slides, 71
polyclonal antibodies, 89
postinjection care of embryos, 131, 142
preparation of DNA for injection, 164
preparative EMSA, 193
preparing zebrafish embryos, 129

primer design for in vivo footprinting, 210
primer extension, 208
probe hydrolysis, 75
probe synthesis, 60, 72
promoter analysis, 133
promoter analysis in zebrafish, 155
promoter analysis in zebrafish experimental considerations, 155
promoter analysis strategy, 156
protection footprinting, 188
proteinase K, 32
pSP64T, 103
purification of DNA for injection, 131

Q

quantification of RNA, 105
quantitative PCR, 54

R

random hexamers, 44
rapid zebrafish WISH protocol, 66
re-using antibodies, 64
reference cDNAs, 52, 55
REMI, 167
removal of an ovary lobe, 116
removing unincorporated NTPs, 181
restriction analysis of ITRs, 172
reverse transcriptase, 44
reverse transcriptase reaction optimization, 48
ribonuclease protection, 29
ribonuclease protection, experimental considerations, 29
ribonuclease protection kits, 39
Ringer's solution, 16
RNA degradation, 30
RNA gel electrophoresis, 106
RNA isolation, 32, 43
RNA metabolism, 112
RNA probe, specific activity, 37
RNA probe, storage, 38
RNA probe purification, 33

RNA probe synthesis, 33
RNA quantification, 4
RNA yields, 44
RNase A, 34
RNase T1, 34
RT-PCR, 41
RT-PCR, experimental considerations, 41

S

SDS-PAGE, 95
secondary antibodies, 87
secreted proteins, 111
sectioning, 72, 81
sectioning wholemount zebrafish, 86
Spemann organizer, 2
STABL2 cells, 170
Steinberg's solution, 139
supershift, 176, 178, 182
SURE2 cells, 172
synthetic RNA, 99

T

TESPA coated slides, 81
thermal cycler, 42
thymidine kinase promoter, 155
tissue specific expression, 168
tissue transplantation, 16
tissue-directed injection, 147
Tm estimation, 49
transcription factors, 175
transfection, 23, 137, 148
transgenic fish, 125
transgenic *Xenopus*, 133, 167
transgenic zebrafish, 157
translation regulation, 112
tRNA, 42
tungsten needle, 11, 17

V

vectors for synthetic RNA, 103
Vent DNA polymerase, 203
ventralized fish embryos, 15, 20

Index

vibratome, 92
vibratome sectioning, 94
viral ITR sequences, 137, 167
vitelline membrane removal, 5
volumes for injection, 119

W

Western blotting, 90, 95
wholemount immunohistochemistry, 81
wholemount in situ hybridization, 57
wholemount staining of animal caps, 12
WISH hybridization mix, 59
WISH of *Xenopus* embryos, 61
WISH of zebrafish embryos, 63

X

Xenopus immunohistochemistry, 89
Xenopus sections, 72
Xenopus suppliers, 120
XMyoDb, 50

Y

yolk cell transplantation, 15, 20

Z

zebrafish immunohistochemistry, 77